The Diary of
Col. John Henry Stover Funk

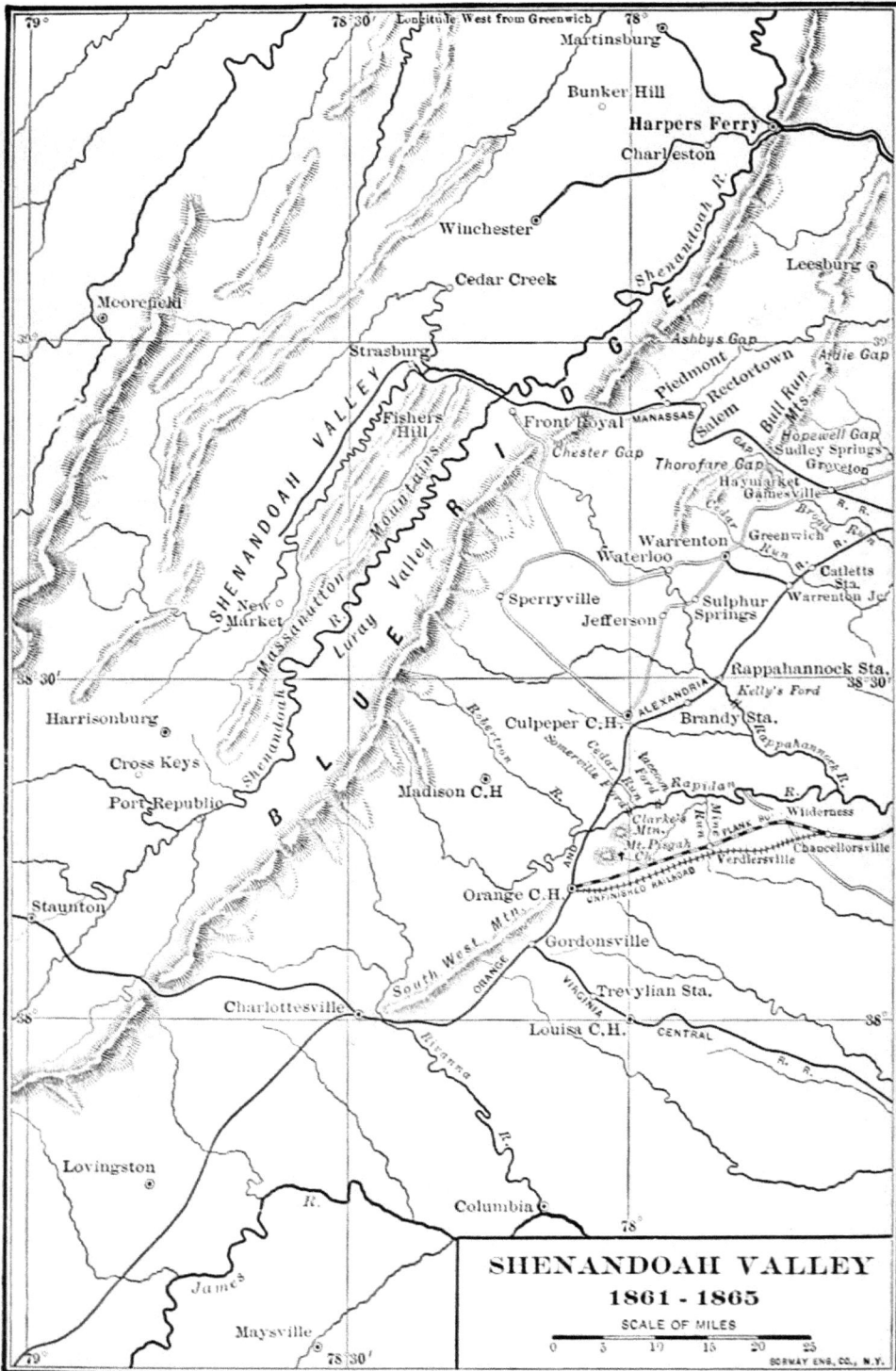

SHENANDOAH VALLEY

1861 - 1865

SCALE OF MILES

0   5   10   15   20   25

map courtesy of loc.gov

THE DIARY OF
COL. JOHN HENRY STOVER FUNK
OF THE
STONEWALL BRIGADE
1861-1862

Joe D. Haines Jr., MD, Editor

SHOTWELL PUBLISHING
COLUMBIA, SOUTH CAROLINA

Produced in the Republic of South Carolina by

SHOTWELL PUBLISHING LLC
Post Office Box 2592
Columbia, So. Carolina 29202
www.ShotwellPublishing.com

ISBN: 978-1-947660-67-0

First Edition

10 9 8 7 6 5 4 3 2 1

TABLE OF CONTENTS

The Editor would like to acknowledge the fine work of Professor Clyde Wilson in the preparation of this diary for publication. Another page has been added to the chronicle of the War for Southern Independence thanks to his able assistance.

PREFACE

The previously unpublished diary of John H.S. Funk is an important source for the first year and a half of the War Between the States in Virginia. Funk was a frontline combat officer—captain, lieutenant-colonel, and colonel of the 5th Virginia Infantry Regiment in the famed Stonewall Brigade. He vividly records, almost day by day, a soldier's life and battles—First Manassas, Jackson's Valley Campaign, the Seven Days, Second Manassas, Cedar Mountain, and others. Funk appears as a real man as he records the miles marched by Jackson's "foot cavalry," first-hand observations of Yankee war crimes against civilians, and the terrible death toll of Confederate fighters.

—Shotwell Publishing

INTRODUCTION

AMERICA'S WAR BETWEEN THE STATES continues to be one of the most studied and hotly debated episodes of American history, even though the last shot was fired over 150 years ago. Tens of thousands of volumes have been published concerning this pivotal event in our past, and each year more books are added to the library. Why is it that America retains a particular fascination with this war? On the surface, it is easy to point to the fact that this was a tragic epic drama, in which brother literally fought against brother, as occurred in my own family.

But below the surface, as one delves deeper into the causes and effects of the war, the realization develops that the War Between the States, more than any other event save the Revolutionary War, defines who we are as a people. The outcome of the war remade our country, destroying state sovereignty forever while establishing an all-powerful federal government that we labor under today.

Historians, scholars and amateurs alike, will endlessly debate the "true" causes of the war, and after years of study I also have my favorite opinions. But that is not the purpose of this book. My purpose is rather to provide what I believe to be a valuable contribution to the study of this significant event in American history: the discovery of a new primary source, namely the diary of Colonel John Henry Stover Funk.

Professional historians know well the value of primary sources, as new eyewitness accounts of an event that occurred over 150 years ago are far and few between. These newfound eyewitness accounts excite considerable interest, especially in providing new insight and confirming or refuting what has been accepted as fact. New material also upsets the status quo, as I discovered, because the experts often become very partial to their particular interpretation as the "truth." The experts are of course correct to question the historical validity and accuracy of new material, as frauds have been perpetrated in the past.

THE DIARY

Although I am not a professional historian, I have researched Colonel Funk's diary over the past seven years and believe that it is the genuine article, and as such a valuable contribution to Civil War literature. The diary, which begins in April, 1861, provides much valuable information concerning the tumultuous early phases of the Civil War. It concludes with the opening of the Maryland campaign in September, 1862. This leads one to believe that this was the first volume of a series of diaries, given that Colonel Funk was not killed until 1864.

The highlight of the diary is Funk's description of how General Thomas J. Jackson secured the most famous military nickname, "Stonewall." Most accounts credit the name "Stonewall" to General Barnard Bee at the Battle of First Manassas. According to Funk however, credit should go to Colonel Francis Bartow, commander of the Eighth Georgia, who in speaking to General P.G.T. Beauregard stated, "Look at Jackson. He stands like a stone wall."

But beyond this discovery, Funk's diary provides an excellent first-hand account of the opening months of Civil War campaigning in Virginia. Funk reveals valuable information concerning battles and military movements, day-to-day activities of his company and regiment including camp life, weather, the suffering of civilians from Northern invasion, health, and other matters. He provides vivid guidance to the heroic marches of Jackson's "foot cavalry."

The preservation of this remarkable diary is something of a miracle. According to collector and artist Alan Curtis of Arkansas, the original diary was purchased by a Mrs. Mickles near Clarksville, Arkansas, at an antique store in 1970. Mrs. Mickles had a great-grandfather who had served in the Virginia cavalry, and she was an avid collector of Civil War memorabilia. Mrs. Mickles showed the diary to Mr. Curtis, who described it as "a russet-colored leather pocket diary common of the 1860s."

Mr. Curtis was eager to read the diary and Mrs. Mickles agreed to have a copy of the diary typed.

According to Mr. Curtis, "Mrs. Mickles had one of her office staff type a copy of the diary for me. There are several misspelled words, but I don't know if it was typed as written by Colonel Funk, or if the errors were made by the young transcriptionist. I remember the writing in the diary being neat and legible. I am sure that this was just one of several diaries that Colonel Funk did because he ran out of pages to write on."*

Mr. Curtis continues, "Also about this time Mrs. Mickle was wanting to attempt to find a Funk family member so that she could give the diary to them. She felt it should be in the family. She had written to historian Bruce Catton regarding how Jackson got his name Stonewall in the diary. She said Mr. Catton felt it interesting, but his remarks were something like, 'When rumor becomes fact, then it becomes history.' I took this to mean that once it's in the history books you can't change it."

Mr. Curtis stated that he eventually lost contact with Mrs. Mickles; however, he related the story to retired Professor Phillip Rutherford of Arkansas Tech in Russellville, Arkansas, who made an attempt to locate the diary years later. Professor Rutherford discovered that Mrs. Mickles had died and that her surviving daughter was unaware of the existence and whereabouts of the diary.

---

* The transcript published here is full of misspellings, whether the work of the diarist or the transcriber is not always clear. Nineteenth century handwriters spelled casually compared to today's practice. Rather than stud every page with corrections, I have left most of the misspellings as they are, understanding that the reader will recognize most of the words and will get used to the fact that Funk used the word "quite" frequently and spelled it "quiet," and that he several times refers to General Turner Ashby, Jackson's cavalry commander in the Valley Campaign, as "Ashley."

In 2004, I made contact with one of the Funk family descendants, Rick Dillman of Grand Haven, Michigan. According to Mr. Dillman, he possessed copies of correspondence between Col. Funk and his family from 1858 until his death in 1864. He claimed that the family had no idea of the existence of the diary, since all of Funk's personal effects were lost following his death at the Battle of Winchester in 1864. According to a Funk family tradition, a document was found in Col. Funk's pocket, recommending his promotion to brigadier general. It lacked only General Robert E. Lee's signature.

Mr. Dillman informed me that Col. Funk's letters had been rediscovered by the family in the mid-1990s. They were contained in an old leather bag with the initials "J.B." carved upon it hanging in the back of a closet. Funk had been a militia commander in Virginia prior to the war and was present at the hanging of abolitionist John Brown and sons in Charleston in 1859. Mr. Dillman suspected that Funk may have been involved in digging up the body of one of Brown's sons, whose corpse turned up in the dissecting lab of Winchester Medical College where Funk was a student at the time.

Mr. Dillman stated that Funk's letters to his family rarely mentioned details about the war, presumably hoping to shield his parents from the carnage that he witnessed. Only once did he write about a close call during a skirmish where he received multiple bullet holes in his coat. Funk's younger brother, Jefferson William Obet Funk, served with him in the 5th Virginia. He was taken prisoner in May, 1864, and became one of the Confederate prisoners that the Union army deliberately exposed to fire at Morris Island—one of the "Immortal 600."

COLONEL FUNK AND THE STONEWALL BRIGADE

The Funks were Mennonites exiled from Switzerland in the late 1600s. They immigrated to America on the ship *Mary Hope* in 1710, with Col. Funk's direct ancestors relocating to the Shenandoah Valley in 1736.

John Henry Stover Funk was born in Winchester, Virginia, on June 28, 1837. Little is known of his early years; however he graduated from Winchester Medical College at the age of 23 on April 23, 1860. Dr. Funk hung out his shingle in the town of Farmington, Marion County, Virginia in August 1860. He practiced there until April 1861, when he enlisted in the Confederate Volunteers at Harper's Ferry.

Curiously, Funk elected not to serve as a surgeon, and was appointed Captain, Company A, 5th Virginia Infantry Regiment, on April 28, 1861. The 5th Virginia became part of the famed Stonewall Brigade. Funk rapidly rose in rank to lieutenant colonel by April 29, 1862, and became colonel commanding the regiment on August 29, 1862.

The 5th Virginia was known as "The Bloody Fifth." It had its origin in a volunteer militia regiment organized in Augusta County on April 13, 1861. When the regiment was accepted into the service of the Confederate States, it consisted of seven companies from Augusta County, two from Winchester and Frederick County, and one from Rockbridge County. The 5th Regiment, commanded by Colonel Kenton Harper, was part of General Thomas J. Jackson's First Brigade. They had their baptism by fire on July 2, 1861, at Falling Waters.

The 5th was outstanding in its performance at First Manassas. Upon the reorganization of the 5th in April, 1862, a band was added to the regiment. It was destined to become the famous Stonewall Brigade Band which still exists by that name today.

As part of the first brigade, which became known as the Stonewall Brigade, the 5th fought through the 1862 Valley Campaign and the Seven Days' battles. The regiment fought at Sharpsburg, Fredericksburg, Chancellorsville, and Second Winchester. At Gettysburg, under Colonel Funk, the 5th Virginia was involved in the desperate fight for Wolfe's Hill.

Thirteen members of the 5th Virginia were placed on the Roll of Honor for gallantry at Payne's Farm on November 27, 1863. The 5th went into winter quarters on the Rapidan River from 1863-64, where the soldiers entertained themselves with theatricals, band concerts and other diversions to break the monotony of camp routine. The most memorable event recorded was the famous snowball battle of March 23, 1864.

The 5th next fought at the Battle of the Wilderness in May, 1864. Brigade casualties were reported as heavy; however the losses the following week at Spotsylvania were disastrous. Funk ended his life as a warrior, mortally wounded at the Battle of Third Winchester and dying two days later on September 19, 1864. He was taken to his parents' home in Winchester to be cared for. He reportedly asked his mother if his wound smelled. She replied that it did, and Funk said that he was going to die.

The remnants of the old Stonewall Brigade were organized with the survivors of thirteen other decimated Virginia units to form William Terry's Brigade, which with John B. Gordon's division, served with Jubal Early in the Shenandoah Valley. After the battle at Cedar Creek, the brigade was sent to the lines at Petersburg, where they fought in the battles at Hatcher's Run and Fort Stedman. Following General Lee's surrender, the 5th Virginia was paroled at Appomattox Court House.

Colonel Funk's diary contains a wealth of information concerning the opening months of the Virginia campaign. Included are descriptions of First and Second Manassas, Jackson's 1862 Valley Campaign and The Seven Days. The early activities of the Stonewall Brigade are particularly well covered and reinforce and supplement standard references such as James Robertson's *Stonewall Brigade*.

It is a pity that the diary abruptly ends on the eve of the Battle of Sharpsburg (Antietam). It would be nice to think that Col. Funk continued his diary until his death in 1864, and that subsequent volumes may one day be discovered.

Naming Stonewall

As mentioned, the highlight of the diary is Funk's eyewitness account of General Jackson's receiving his nickname, "Stonewall." Most accounts credit General Barnard Bee as the source, however there are numerous versions of what Bee said. Funk's report is the only eyewitness account and provides the startling revelation that Col. Francis Bartow, not General Bee, made the famous Stonewall reference. In his diary entry of July 21, 1861, chronicling his experience at First Manassas, Funk wrote that he heard Col. Francis Bartow, whose command was adjacent to his, say. "Look at Jackson. He stands like a stone wall."

Both Bee and Bartow were killed during the battle and left no record of the matter. The diary of Funk is here published for the first time, never before being available to historians.

James I. Robertson, the authority on Stonewall Jackson, writes that Bee's precise words will never be known with certainty. The first published account was a correspondent's story in the *Charleston Mercury*, July 25,1861, reprinted verbatim in the *Richmond Daily Dispatch*, July 29,1861, and the *Lexington Gazette*, Aug. 15, 1861. Robertson notes eight articles, books, and speeches before 1900 that carried the Bee story with slight variations.

One important quoted rendition has the phrase "stone wall" directed at Jackson's men rather than to their commander. In this case, Bee is thought to have said: "Look at Jackson's brigade standing like a stone wall! Rally on the Virginians!" One could argue that this is the most realistic interpretation, since a line of men bears more resemblance to a wall than would an individual—even Jackson. Jackson always said that the name applied to the brigade, not to himself.

This reference to the Virginia brigade also originated immediately after the battle. It remained popular for a quarter-century following the war. Among the first men to present this version was Brig. Gen. (later Episcopal bishop) Ellison Capers, a fellow South Carolinian and good friend of Bee's. Jackson's Surgeon Hunter McGuire led a host of Virginians by combining the two versions and having Bee shout: "There stands Jackson like a stone wall! Rally behind the Virginians!"

At least three reports of the incident have twisted the facts out of context. The most preposterous account came from Jackson's first biographer, Markinfield Addey, who stated that Jackson himself coined the phrase "stone wall" while assuring Beauregard that his men would hold Henry Hill. Another variant version of the event is given in Brig. Gen. E. Porter Alexander's well-known memoirs. In this version, General Joseph E. Johnston, army commander, is reported to have turned to .Alexander that afternoon and praised the 4th Virginia alone for standing "like a stone wall." Henry Kyd Douglas, in his *I Rode with Stonewall,* wrote that Bee had "the excitement and mortification of an untried but heroic soldier." But Bee was a veteran of the Mexican War and other service.

Exactly what Bee meant in his war cry has received two negative interpretations. This viewpoint began with two staff officers: Maj. (later Maj. Gen.) W.H. Chase Whiting, whose resentment of Jackson at Winchester in the spring escalated in time to open dislike, and Maj. Thomas G. Rhett, South Carolinian. These two men, with Whiting by far the more outspoken, told another South Carolina officer, John Cheves Haskell, that Jackson refused to advance to Bee's relief. "In a passionate expression of anger [Bee] denounced him for standing like a stone wall and allowing them to be sacrificed." No established facts support l that statement. One of Bee's own aides, William P. Shingler, wrote a little-known account that refutes all negative overtones placed on the origin of the nickname. See *New Eclectic Magazine,* 4 (1869): 745-46.

Jackson's brother-in-law Gen. D. Harvey Hill dismissed the "stone wall" story with the remark that the name was unsuited to Jackson, who "was ever in motion, sweeping like an eagle on his prey." *Century Magazine,* 47 (1893-94): 623.

More than fifty years ago, Douglas Southall Freeman discussed the pros and cons relative to the "Stonewall" name in Lee's *Lieutenants: A Study in Command* (New York, 1944), I:733-34. His conclusions remain sound. The story of "Stonewall" Jackson's exploits were circulating joyfully in Richmond three days after the battle. A South Carolina newspaperman, who was the first to report what Bee shouted, would have never mentioned the story—or been as laudatory—if the incident had not had positive overtones. The enthusiasm attendant to those early days of the war gives the action the appearance of probability. Freeman concluded that it was up to anyone not believing the traditional origins of the nickname to offer convincing evidence to the contrary.

The discovery of Colonel Funk's diary meets Freeman's criteria for providing evidence disproving the traditional origins of the nickname, "Stonewall." Funk's account is the only known eye-witness account. Since Funk died in 1864 before the war ended, his writings were not subject to editing, embellishment or reinterpretation, as occurred with so many other versions. After Bee's wounding, Bartow reported to Beauregard and the two officers conferred. Beauregard ordered Bartow to take his men to Jackson's left. Once Bartow had his men in line, he was killed. Thus a plausible scenario existed for Funk to have heard Bartow's comments. In contrast, General Bee's supposed statement was made after he returned from conferring with Jackson. When Bee rejoined his men, Capt. Thomas Goldsby, quoted Bee as saying, "This is all of my brigade that I can find—will you follow me back to where the firing is going on?" Goldsby described the scene a few days after the battle, saying his men responded, "To the death!" Goldsby mentions nothing about the famous nickname, which seems curious, given how inspiring such a statement would have been. Other soldiers present supposedly told a correspondent a day or two later that it was Bee who made the comment, and a legend was born.

The Funk diary is thus in my opinion a valuable contribution to Civil War literature. It is unfortunate that the original diary has been lost, since it could have provided confirmation of the typescript.

Many professional historians will likely discount the authenticity of the typescript and insist on authentication before accepting this new information. Is it possible that the typescript is a fabrication? Perhaps, but if so, it would certainly be an exceedingly elaborate hoax, and to what purpose?

—**Joe D. Haines, Jr., M.D.**

A diary of the chances and hardship of the service, kept by J. H. S. Funk, Col., 5th Va. Inf., Stone Wall Brigade.

Harpers Ferry, Virginia: April 27, 1862

My object in noting the hardships and pleasures of my tour in this struggle for freedom and right is purely for my own gratification, should I survive. If not, for the benefit of my friends who are interested.

If thare be in this hasty record of my carreer as a soldier showed fair efforts as an officer, an act worthy of admiration and entitled to the applause of my brothers in arms, I record it to the honor of a pious mother and kind father, to them it is due from the influence of an early tuition.

My only regret is that I am unable to reflect more credit to them. Be that as it may, I am determined to comply with their desire which is to service my country faithful and never dishoner thair name.

Respectfully,

J. H. S. Funk, Colonel
5th Va. Inf.
Stonewall Brigade

APRIL 1861

April 28, 1861

I left Farmington today at 2:00 P.M. where I had located and practiced medicine for fourteen months with unexpected success for join my company at Harpers Ferry.[1]

It was with some regret that I left the pleasant village of Farmington and the many friends I had won in my sojourn. Fortune had thrown me among them, a total stranger and a poor boy, just from college, without a practicable knowledge of my profession and without money, for after paying for my stock of medicine, I had barely enough to pay a month's board. But with prudence becoming a gentleman and successful as a physician, I had won the esteem, friendship, goodwill, and confidence of the citizens. My departure seemed as unpleasant to them as it did to me. In taking leave of them, they sent with me their best wishes, after they had exhausted every entreaty to keep me from going, but I owed my country a duty which I was honor bound to perform. In many instances tears had been shed.

It was there that I found one which proved to be more than a friend to the stranger. When worn out with the duties of the day and worried with the society of men, I sought her presence, who I found ever ready and willing to sympathize and interest me, with an

---

1 Harper's Ferry. Site of a federal arsenal in Virginia where a militant abolitionist, John Brown, led a group of 18 men on a raid October 16,1859. Brown's goal was to confiscate weapons, arm slaves and incite a slave rebellion. A contingent of U.S. Marines commanded by Colonel Robert E. Lee put down the rebellion. Brown was captured, tried and sentenced to hang. Concerned about attempts to rescue Brown, Governor Wise of Virginia ordered militia units and cadets from the Virginia Military Institute out to preserve order. VMI professor Thomas J. Jackson directed 21 cadets who manned two howitzers, along with 64 cadets forming two companies of infantry.

intelligent and well stored mind, which is rare in this locality. How painful it was to separate under the existing circumstances and be deprived of the refining influence.

After receiving several handsome boquets from the ladies for my company, I mounted the train bound east. We were soon under way and it is impossible to describe my feelings as moments distanced me from the delightful village which I respected as my adopted home. I was somewhat consoled by nearing the city of my nativity [Winchester] and the home of my family, which I had been absent form for more than a year. I fancied too the welcome of the friends of earlier days. Had the circumstances been different, it would have been but pleasure, but the political condition of the country was such, that the best and most learned men of the nation feared the issue. The storm grew daily more threatening, while the feelings which united the south with the north became less amicable. Fort Sumpter had already surrendered, the government buildings at Harpers Ferry had been burned, and enemy troops held possession of the place, expecting to meet the enemy in a hostile conflict ere many days. A revolution was enevitable and God only knew where it would end when once fairly begun. The thought of the distracted condition of my country was enough to make the going sad. The uncertainty of war, too, caused the thought frequently to arise that perhaps I should never again see the friends I had just left. I would banish the thoughts like most unexperienced young men, with the romantic idea of a woman's laurels and brilliant achievements of southern chivelery and valor in a just and righteous cause.

In half an hour the train stopped at Fairmont, the county seat of Marion, situated on top of the Mononghala river, twenty seven miles from its mouth. Upon the opposite bank was Palatine, a village of some fifteen hundred inhabitants. These towns were connected by a suspension bridge singular too, that so close as these towns were, only separated by a river of three hundred yards in width, yet they were entirely opposite in sentiment. The citizens of Fairmont were with the south and justifying her present course, while Paletine condemned and clung to the union with the same verocity that a yankee clings to a sixpence. A number of my acquaintances knowing my intention, had already collected me to bid me God's speed, promising that they would follow in a few days.

Again we were in motion and after an hours ride we reached Grafton, the junction of the West Virginia railroad, twenty miles east of Fairmont. Col. Robertson[2] was addressing a crowd who had just raised to breeze the stars and bars already made famous by southern courage. The populace seemed somewhat divided. A gentleman on the train noticing it and judging from the sarcasm of his remark that he had but little interest in the south. He said, "You must have more unity of sentiment than in the south, or they will fly like dogs before the discipline of a united north." His remarks made me a little uneasy for a while for I was ignorant of the unity or purpose of either, having remained quiet at home, and only judged on the sincerety of the south by the nature of wrongs imposed. Could it be that a few fanatics had set on fire the ignorant population of the south, but I had too much respect for the intelligence of the south and the noble resistance they had made against a sectional and unwarrantable wrong to permit my mind to be troubled long. The second thought sufficed.

After changing engines and receiving on board a few more passengers, we started up the glides, we stopped several times to get wood and water. I noticed at several stations along the road that the people assembled together at the call of Brown and Carlile to protest against the action of our State with a view, too, of appointing delegates to the convention which they proposed holding at Wheeling. They were silently attended and principally by such as depended solely upon the railroad for existance. But progress was

---

2 Colonel Robertson. Colonel Beverly H. Robertson, later promoted to brigadier general and commander of a cavalry brigade in Stonewall Jackson's army. 1849 West Point graduate with cavalry experience against the Indians in the West. Known as a strict disciplinarian, he was popular neither with his superiors nor his men. Promoted to brigadier general June 9, 1862. Led troops in Shenandoah Valley and Second Manassas. Performed poorly at Cedar Mountain. Transferred by Jackson. His troops were routed at Brandy Station. Failed to follow orders at Gettysburg. Transferred to South Carolina, defended Charleston for a month and surrendered.

somewhat slower since we had left Grafton, for it was upgrade to Cookland. We soon started into Kingwood Tunnell, nearly a mile in length, then we struck the Cheat River trestling, celebrated for the neatness of its structure and the grandeur of the surrounding scenery. As the train moved slowly along for the benefit of those on board, one would become enchanted with the sublimity of the view. The train seemed as if almost suspended in the air by the network of iron beneath while we could look for four hundred feet into the abyss below. One will always forget the shortness of time, we were permitted to remain there. It would take a more abler pen than mine to give the remotest idea of its grandeur.

The whistle of the train told us we were entering Rowlesburg, a village of three hundred inhabitants, situated in the west bank of Cheat River in Preston County. Geo. E. Towens met me at the cars and expressed his wishes of entering our service and desired my assistance in getting him a commission.

As soon as watered, we again commenced ascending in the Alleghaney. In the dark, we stopped for supper at Oakland, which is a resort for visitors from some of the more southern states, being situated on top of the mountain. The number of visitors and the beauty of the town gave it a striking appearance to the traveler who had become tired and warried by the distance they had traveled and the confines of a crowded car.

After washing the dust from our face, we entered the dining room. I have fancied from the manner in which the contents of the plates within reach had dissappeared, the steady tread of the servants and the dejected countenance of the landlord as he watched me as I complied with the demands of a whetted appetite, that I had done more justice to the satisfaction of my hunger than I ever would to the justice of my cause in the future before the enemy. Supper was soon dispatched and the husky voice of the conductor, "All aboard" summoned us to the cars. Scarcely seated, when the train pulled out. It had become dark in the course of a few hours.

We were again halted at Cumberland, an enthusiastic union crowd had surrounded the cars as soon as halted, interrogating every passenger as he stepped from the train on the platform with, "Are you seccesh or union." Upon telling them that I was a southerner, excited their indignation following close at my heels until I entered the hotel. I began to fear that they would detain me, but fortunately and to my surprise, all passed off well.

April 29th

We passed through Piemont, meeting several of my acquaintances. At 11 o'clock P.M., we stopped at Martinsburg,[3] the county town of Berkley, containing some four thousand inhabitants. Here several soldiers came aboard, being the first uniform I had seen, was quite exciting, giving everything the appearance of "glorious war." They appreciated their becoming dignity of military gentlemen, making everything show off to the best advantage, repeating to the eager passengers the miraculous adventure of the exposed condition which they had endured in the beginning as they thought of the second John Brown War. It is to be feared that this pen will be constrained to give items of war in a more real sense before the drive will cease to arouse my fellow countrymen to duty, We pray that something may entervene to prevent the issue of blood.

April 29th at 2 o'clock P.M.

We were stopped within a mile of the ferry by a police guard, whose instructions were to search the train. An officer with a plume epaulettes and dangling sword, with an air becoming that of a warrior, came aboard accompanied by half a dozen men, with glittering guns and fixed bayonets. They demanded of those on board some assurance as to their loyalty to the south[;] those who could not were immediately taken under charge to the guard house. Having my commission with me, all passed of quietly and I was permitted to roam at large.

---

3 Martinsburg, (West) Virginia. County seat of Berkeley County, located approximately 20 miles northwest of Harper's Ferry and 20 miles north of Winchester. (see map)

I admired the military dignity of the officer who was on board. He did his duty with as much grace as Marshall Nay[4] would have in receiving the sword of a surrendering general.

Several had no papers to show what or who they were. At once they were taken under charge of the guard for further investigation. I at once sought the whereabouts of my company which I had learned was there, and quartered in the armory. I attempted to reach the gate, but every fifteen steps I was halted by a green sentinal by, "Who comes thare" with his bayonet pointed directly at me, and I could see from his countenance that if I dared to step forward, my fate was certain. Upon explanation, together with the display of my commission which I kept constantly in my hand for ready use, until I blundered upon an Irishman on post, who in spite of my credentials and assurances, called the sargent of the guards, who made his appearance. I could do more with him than I could do with the sentinal. Without further ceremony, I was escorted between two desperate looking soldiers to the guard house. Here is a blunder, I thought, disgraced, confined in the guard house. Already immortalized, my anticipations of future glory and distinction gave way to feelings of remorse. Upon arriving at the guard house, I was turned over to the officer of the guard, who examined my papers, and to my great relief dismissed me.

One of my men by the name of Hugh Bar and a playmate happened to be on guard, he recognized me and at once conducted me to the company. They were asleep. I soon aroused them. They were highly delighted upon my arrival. Lt. A. S. Maskell was in command, having just returned from college.

April 30th

The rest of the night was dedicated to calling over by-gone days and giving them an account of my adventures during my absence. At 9 A.M. I reported to Col. Morse[5] for duty. He was in command of the 6th Va. Regt. of Infantry. Everything had a military appearance,

---

4 Marshal Ney. One of Napoleon Bonaparte's Marshals.

5 Colonel Morse. Perhaps this is a falsely transcribed word. Possibly Funk means William Mahone, at this time colonel of the 6th Virginia. There was no Col. Morse.

the streets were crowded with dashing officers and handsomely uniformed men. While the ear was deafened with mortal music. Col. Morse's staff at that time consisted of Wm. R. Demey, Lt. Col.; Majors Washington and Riley; Adgt. E.R. Smith; Sergt. Major Ed McGuire; Surgeons J. P. Smith; Hunter McGuire[6] and Robert Houston; Qt. Master L. E. Crum; Commissary Assistant Qr. Master Doc Jolliff; Assistant Commissary Col. Singleton; 2nd Genl. Capt. Clark and about twelve orderlies all occupying the same house.

I went to work at once, got clothing and shoes for the men. I found the company somewhat neglected, commencing to try and reorganize it, if possible. Being Sunday, I gave to each a sprig from the boquet and it was worn by them to church.

In the afternoon, I made application to visit Winchester, which was granted. At 3 o'clock P.M. I took the train to Winchester. The train was heavily laden and it ran very slow. We reached the place at sunset. Father, Jenney and Nellie were there to receive me. I perceived that my absence had made some change for I passed through my family without being recognized. I was not the pale faced student, I had become hardened by exercise and gotten fleshier. On my way home, I met with Andrew, one of my old friends who welcomed me back to the town of my nativity. As I neared the abode of sunnier days, I quickened my pace for I was anxious to see that pious mother of mine to whom I had been disobedient. The separation seemed to make me appreciate her real worth. I was anxious to acknowledge my wrongs and seek her forgiveness.

Nellie and I attended the Methodist Church South and on our return, mother and I sat until a late hour in the night talking over the changes which had transpired in my absence. I told her of my success, which was gratifying to her. We at times would talk until we were both in tears.

---

6 Hunter McGuire. Staff surgeon for Stonewall Jackson. Jackson's personal physician throughout the war. Amputated Jackson's left arm following his wounding at Chancellorsville. Attended Jackson until his death ten days later, probably due to a pulmonary embolism rather than the traditional diagnosis of pneumonia. Following the war, Maguire was elected president of the American Medical Association and American Surgical Association.

April 31st

Early this morning I was up and visiting my friends and hunting up the delinquents of my company. This was the first I saw the gift from Harry to Virginnia. At 3 o'clock I returned to the Ferry. On the train was Col. T. J. Jackson,[7] who was to take command of the troops which were now commanded by Kenton Harper,[8] Major Genl. Va: Militia. I found that my company improved somewhat on my return.

---

7 Col. T. J. Jackson. Thomas J. "Stonewall" Jackson, b. January 21, 1824, Clarksburg, (West) Virginia. Graduated West Point, 1846, served in the Mexican War 1846-47, received two brevets for gallantry. Resigned army commission in 1852 to take professorship at Virginia Military Institute where he remained until the outbreak of war in 1861. Appointed Colonel of Confederate Volunteers April, 1861, Brigadier General June 17, 1861. Won fame at Manassas, July 21, 1861. Promoted to Major General October, 1861 and commander of all forces in the Shenandoah Valley. Won fame in the Valley Campaign, winning victories at Front Royal, Winchester, Cross Keys and Port Republic. Joined General Robert E. Lee in driving McClellan from the Peninsula at Seven Days Battles. Defeated General Pope at Groveton and Second Manassas. Won distinction at Battle of Antietam. Promoted to Lieutenant General October, 1862. Commanded right flank at Fredericksburg. Won stunning victory against General Joe Hooker at Chancellorsville, where he was wounded and subsequently died of complications, May 10, 1863.

8 Kenton Harper. b. Chambersburg, PA., 1801. Staunton newspaper editor and banker who had been a volunteer in the Mexican War and served a term in the Virginia House. Was 60-years-old at the start of the war. Militia general who was reduced in rank by the state convention to Colonel of Funk's 5th Virginia Volunteers. Fought at Battle of First Manassas. Resigned commission in August, 1861, when General Jackson refused his request for a furlough to visit his dying wife. Served as Colonel of reserves at Battles of Piedmont, June 5th 1864, and Waynesboro, March 2,1865. Died 1867.

MAY 1861

Tuesday, May 1st, 1861

Commandants of companies were ordered to report to Col. Jackson. We were required to report the conditions and number of our companies. Upon being introduced to the col. I found him to be a very pleasant gentleman. Regiments were at once reorganized. The militia officers who had enjoyed the pompt of war, were busy in packing up, preparing to leave. It was amusing to see the stir that was kicked up upon the removal of these officers. A great deal of dissatisfaction was expressed upon the actions of the state convention[1] then in session in the City of Richmond. I regretted very much to lose Col. Morse.

Wednesday, May 2, 1861

The number of troops now at this place numbered about 23 hundred, being unfamiliar with such affairs. We imagined that we could have successfully engaged and whipped the combined forces of the enemy. The troops were principally those who had a short while before been summoned to the same place by the introduction of John Brown.

Every officer and man alike showed off to the best advantage by always appearing in their best suit, ready to be introduced or to introduce his friends, which was very popular among those who had but little to do, and thought this was of short duration, and that they

---

1 Actions of the state convention. According to historian James Robertson, "On April 27, 1861, Virginia Governor Letcher informed army garrisons throughout the state that all militia officers above the rank of captain were relieved from duty upon the arrival of a duly designated army officer. Most of these militia commissions were issued years earlier, when Virginia's military system existed in name rather than in fact. Letcher's directive bore down especially hard on the Harper's Ferry officers."

would make the best of it. On drills and dress parade especially, the men were required to appear in regimentals. If one was so unfortunate as to not havve his boots blacked, he was censored, receiving a severe repremand and placed on double duty. The company that appeared in the finest uniform was considered the best appearances, was the criterion by which our valor was judged. This only stimulated the pride of the vain. Plumes epuletts and swords were in abundance.

Thursday, May 3

In the reorganization of the regiments, my company was thrown into the 3rs, formerly commanded by Col. W. S. Baylor of Augusta[2] County, who was superceded by Col. Jackson's order. Being the senior officer, I was placed in command until the convention could appoint a col. I found much dissatisfaction in this regiment in consequence of the removel of their col, which they appreciated as a gentleman and officer. The regiment was composed of the following companies: "Marion Rifles," "Mountain Rifles," commanded by R. L. Doyle: "Bunker Hill Rifles" James Gilkerson: West Augusta Guards, Lt. Walters: Southern Guards. H. J. Williams: "Augusta Rifles" Jas. Anthrem: "Southern Grays" J. W. Newton. "Augusta Grays" Stewart Crawford: "Rockbridge Rifles" Samuel Letcher: and the "Morgan Continentals" Captain Avis.

---

2 Colonel William Smith Hanger Baylor. Born Staunton, Virginia, April 7, 1831, Common-wealth Attorney for Staunton, 1857. Organized a company called West Augusta Guards and reported to Harper's Ferry. Relieved from command and reduced to captain by ordinance of state convention. Esteemed by General Jackson for remaining at Harper's Ferry to assist in transition from militia to Confederate command. Promoted to major, May 1, 1861, Lt. Colonel, June 13, 1861. Ordered to duty with General Jackson, Winchester, Va., November 1861. Promoted to colonel, 5th Regiment, Va. Volunteers, April 21,1862. Commanding First Brigade at Second Manassas when he was killed August 30, 1862.

Friday, May 4th

It was with some delicacy that I took charge of this regiment, having no experience whatever and a very young man, while the other officers were much my senior in years. To work I went, appointing James Bumgarder, Jr., adjutant, establishing headquarters in Boliver at Mrs. Hartshorn. All the companies of this command, with the exception of mine and Captain Gilkeson, were quartered in Boliver.

Saturday, May 5th

We were ordered to parade in the armory yard for inspection. At 10 o'clock all the forces were collected in the yard. After going through an informal inspection, we were dismissed. The weather has been quiet fair for several days. The enemy has been reported as advancing, quiet a commotion in camp. Ammunition has been issued, officers have been making stirring and patriotic appeals to their men. Men are ordered not to leave their camps and must sleep on their arms. Additional sentinals were posted with the most rigid instructions as regards to vigilance.

Sunday, May 6th

I attended church this morning. Father came down to see me. In the afternoon, we visited Maryland side of the river and the different incampments. It was quiet a sight to those who had never seen the display of military. The men were as they thought, anxious for an engagement with the military. Imagining that they could kill scores of them, I must confess, though willing to serve my country that I was satisfied. I did not desire to see the issue of blood and I could not see the glory of a war that would deluge the country with blood and destroy the happiness of a free people. I fear that their desire will yet be satisfied.

Monday, 7th

In compliance with an order from Col. Jackson, I sent Captains [sic] Company on the Maryland Heights to build block houses. I ordered the Marion Rifles and the Bunker Hill Rifles to move their quarters to Boliver in order to concentrate the forces.

Tuesday, 8th

Col. Kenton Harper, Lt. Col. Wm. H. Harman[3] and Major W. S. H. Baylor arrived this morning, have been appointed as field officer of our regiment. Nothing unusual occurring. Everything quiet.

Thursday, 9th

All the company officers were called together and taken to the parade ground and instructed that in case of an alarm, we would assemble upon this ground. Each company was inspected and cartridge boxes filled. Men slept upon their arms. No enemy in 30 mi. Three hundred daring cavalry could have scared us out of town.

Friday, 10th

All has become quiet this morning. A scout was made this morning by Captain Patrick, but no enemy could be found this side of Carlile, Penn. These scares are of frequent occurrences. The sentinals are posted just on the edge of town, the enemy could approach with all ease, but they are as young in war as we and no doubt the authorities in Washington fear our forces from this point. Every day various rumors will float around. It has been reported in Winchester time and again that the enemy has engaged us. The provost guard at that place was in readiness once to move off, but we imagined that this place was impregnable.

---

3 William Henry Harman. Born Waynesboro, Va., February 17, 1828. Volunteer in the Mexican War. Commonwealth's Attorney for Augusta County, 1851-1861. Commissioned Lt. Colonel, May 7,1861, 5th Regiment Va. Infantry. Promoted to Colonel, September 11, 1861 vice Kenton Harper who resigned. Dropped April 21,1862. Volunteer aide-de-camp to General Edward Johnson, May 17, 1862. Colonel of reserves at Battle of Piedmont, June 5, 1864. Killed March 2, 1865 at Gallagher's Mill.

Saturday. 11th

The company received a handsome box from Mrs. Harry Kemp of Winchester, the first of the kind received from the ladies. This furnished us with abundant supplies of luxeries for a week. A committee was appointed to draft resolutions of thanks, expressive of our gratitude for her kindness. I was authorized to wait upon her.

Sunday, 12th

Received several substitutes[4] today into my company. Attended devine services this morning. A beautiful day, roses in full bloom. Several of our friends from Winchester came to see us. Everything is perfectly quiet, no talk of the Yankees today. Our dress parade was well attended this evening by the ladies. Our regiment looks very well. Each company is handsomely uniformed.

Monday, 13th

A day never to be forgotten by all who were with us today. We received orders this evening to march immediately to Sheppardstown, that the enemy have attacked the two companies from our regiment there on duty. We were soon in line and moved off. I never saw men more anxious to engage the enemy. It was a beautiful evening. Everything moved off pleasant. Dixie was in the air. We had not proceeded far before we were caught in a dreadful hail storm. The hail stones were as large as hickory nuts. We were unsheltered and they cut the blood out of many of the men. Had it continued for five minutes longer, I believe it would have killed many of the men. The roads were knee deep in mud. We reached Sheppardstown about 8 o'clock at nite. The citizens had heard we were coming and had prepared for our arrival. No enemy were near. The citizens had heard nothing about it. A few drunken cavalry had reported the approach of the Federals. After disposing of my men, getting them comfortably quartered among the citizens, I sought quarters for myself. Lt. Burke and myself were requested to stop at Dr. Taylor's. Where we were kindly treated and remained for the night. Distance 3 miles. This is our first march. A fine initiation.

---

4 Substitutes. Apparently men who were privately paid to serve in the place of others.

Tuesday, 14th

After collecting my men together, we started back through the mud. It was a very weary march, not accustomed to this part of soldiering. Late in the afternoon, we reached the Ferry. The 2nd regiment had returned the evening before. Distance 16 miles.

Wednesday, 15th

Lt. L. J. Fletcher was ordered to Winchester on recruiting service and to collect all the deserters from this company. The 1st Kentucky Regiment arrived here this evening dressed in jeans, looked well. I never heard such yelling in my life.

Thursday, 16th

Two Mississippi regiments arrived, each man was armed with a large knife swinging to his side beside his rifle, expecting to plunge into it hand to hand. They encamped in rear of us in tents, making quiet a handsome light after night when each tent was illuminated. Several times I walked out with the ladies to see the encampment and to listen to the fine music from their bands.

Friday, 17th

A great many accidents are consequently occurring on account of amount of pistols carried by the men. The commander was constrained to issue an order prohibiting the men from carrying side arms, an order of great benefit.

Saturday. 18

Today the first Court Martial in the Confederate States assembled. The Boomeranges, a company from Winchester, commanded by Captain Harry Sherer, arrived. They were handsomely uniformed, making a fine appearance. They were assigned to the 13th regiment, commanded by Col. A. P. Hill,[5] formerly commanded by Col. Williams.

Sunday, 19th

I marched the company to church on Union Street in Boliver. In the evening went after cherries. I find that the men are coming more demoralized and careless about the Sabboth.

Monday, 20th

Our regiment was changed from the 3rd to the 5th. Beautiful weather. The men are in fine spirits. Several other regiments arrived today from the south. We frequently hear of the Yankees approaching. As usual, turns out to be false.

Tuesday, 21st

A general excitement throughout the city. The Provost Marshall had searched the town and emptied all the whiskey he could find, which greatly enraged that portion of the populace who had lost by the operation and had not yet become accustomed to martial law, which differs materially from the civil law in many respects. Detailed as officer of the day, the first time I discharged those duties, I made the usual visits to the sentinals. Many of them had, like myself, had but little experience.

---

5 Colonel A. P. Hill. Born November 9, 1825, Culpepper, Va. Graduated West Point in 1847. Fought in Mexican War and Third Seminole War. In 1861 appointed Colonel, 13th Virginia Infantry. Brigadier General February, 1862. Major General May 26, 1862. Lieutenant General May 24, 1863. Fought at Williamsburg, Mechanicsville, Seven Days, Gaines' Mill, Frayser's Mountain, Second Manassas, Antietam, Fredericksburg, Chancellorsville, Gettysburg, Wilderness, and killed in action at Petersburg, April 2, 1865.

Wednesday, 22nd

Appointed a commissioner to superintend the voting on the ratification for the ordinance of Secession. Reported to Col. Jackson for instructions and was sworn. We were ordered to take a vote today. Tomorrow was the day for the election. The only reason I can see for it is precautionary, fearing that the enemy, who are aware of it, will perhaps try to frustrate it by an attack. Each company was brought up and as their names were called, they voted either in affirmative or negative. I was glad to find that but six opposed it, but truly sorry that five of those were from my company. I will refrain from giving their name of same for I know in the future they will regret it, although I shall mention the enitials. M.S.B., H.D., D.S., W.A.Y., and J.G. Everything passed off quietly.

Thursday, 23rd

The vote was retaken today. Only four cast against the ordinance. I heard that Fredrick County had polled two hundred and seventy three against it. I was surprised to find so little opposition.

Friday, 24th

Commenced throwing up fortifications about a mile south of the town. Several large pieces were mounted near headquarters. One was placed on Maryland Heights. The Kentucky regiment was stationed there on picket. They set the mountain on fire and killed all the hogs in the neighborhood and committed various depradations upon the citizens.

Saturday, 25th

Two companies was sent from our regiment under charge of Major Baylor to Bath[6] to collect and bring away some guns which were at that place, which excited the citizens very much. Several men shot at the stars and stripes which was floating at that place. No resistance was offered against the troops.

Sunday, 26th

All military operations suspended. Preaching on our parade ground. Nothing worthy of note occurring.

Monday, 27th

We drill some four times a day, which keeps the men busily employed.

Tuesday, 28th

Received a letter from Farmington,[7] the first time I had heard from my friends since I left. The secessionists of Fairmont and the Union men from Morgantown had quite a row. Fairmonters claimed the field.

Wednesday, 29th

Private Jacob Hilliard of my company was accidentally wounded in the thigh, created some considerable excitement among the medical corps, being the first wounded man in this war.

---

6 Bath. Also known as Berkeley Springs, (West) Virginia. Located about 20 miles northwest of Martinsburg.

7 Farmington. Town where Capt. Funk was practicing medicine prior to the War.

Thursday, 30th

Several Alabama regiments arrived. Each arrival added confidence to the troops. Commenced searching the houses in quest of arms. The citizens had carried off a number of arms before our troops took possession of the Ferry the nite the buildings were burned, and in order to secure them, an order had been issued to search for them. It was amusing in many instances to see with what authority some of the lieutenants in charge of a squad detailed for this purpose proceeded. Had a stranger have been present and seen them approach a private dwelling and demand an entrance, he would have thought him to be the commander of the Forces. Many of the citizens were insulted by these brainless officers who are now in new spere. Great allowances must be made for these men. They are unacquainted with the service and all they know in regard to requisite duty of a soldier is in their experience in parading on a sunny day in some town.

Friday, 31st

Brigades were formed. The 1st Va. Brigade under the command of T. J. Jackson, composed of 2nd Va. Infantry Col. Allen, 4th Va. Infantry Col. Preston,[8] 5th Va. Infantry Col. Kenton Harper and 27th Va. Infantry Col. Gordon.

---

8 Col. Preston. John Thomas Lewis Preston. One of the founders of Virginia Military Institute and influential citizen of Lexington, Friend and colleague of General Jackson prior to the War.

JUNE 1861

Saturday, June 1st, 1861

All the companies were mustered into the service of the Confederate States for one year. Some few of the men objected to being mustered in for that time, but they were few. Many of us looked forward still with some hopes of a peaceful adjustment to the difficulties. The sentiment of the citizens of this portion of the state is somewhat divided. Not in sympathy with the north, but in view of an amicable settlement, every day widens the rupture. A number of Marylanders[1] are daily arriving and entering our service.

Sunday, June 2nd

A lovely day. After a usual Sabbath morning inspection, we attended church, spending the remainder of the day in presence of the ladies.

Monday, June 3rd

Captain Letcher[2] was ordered to burn the railroad bridge crossing the opequary[3] between this place and Martinburg. Affairs begin to present a frightful appearance. Troops are constantly arriving at

---

1 Marylanders. Referring to the First Maryland Regiment.

2 Captain Letcher. Brother of Va. Governor John Letcher. Samuel Houston Letcher, born 1828. Graduated Washington College, 1848. Lawyer by profession. Enlisted 1861, Capt., Co. B, 5th Virginia Infantry. Transferred with company to 4th and 27th regiments. Appointed Lt. Col., 58th Regiment Va. Inf., October, 1861, Col., May, 1862. Died Lexington, 11/10/1868.

3 Opequary. Refers to the railroad bridge between Harper's Ferry and Martinburg. The bridge crosses the Opequon Creek.

Washington, while Genl. Patterson[4] is concentrating his forces at Carlile. His pickets extend to Williamsport. I fear hostilities will soon begin. Our troops are confident of success.

Tuesday, June 4th

We received our supply of new ammunition today. (Minnie Rifles) The measles and mumps is playing havoc with the southern troops. Our hospitals are filled. It is also rumored that the small pox has gotten among the patients in the hospital.

Wednesday, June 5th

Received a letter from Farmington in which it states the cruelties perpetrated by the U. S. Troops. I regret it very much. That portion of our state has not been permitted to speak for herself.[5]

Thursday, June 6th

Rumored that the enemy are advancing in force. Everything is alive again and in confusion. Ammunition reissued further instructions given to the sentinals as regards their vigilance and the troops are ordered to lay upon their arms.

Friday, June 7th

Commandants of companies were ordered to report to the headquarters of their sergeant where we were notified on the following morning. The revelle was to be sounded at 4 o'clock. We were until a late hour in the night making preparations.

---

4 General Patterson. Robert Patterson was a Union general who was a veteran of the War of 1812 as Colonel of Pennsylvania Militia. He was 69 years old when he mustered for Civil War service. Relieved from service by General Winfield Scott July 27, 1861, when he failed to prevent Confederate forces in the Shenandoah Valley from linking up with forces at Manassas.

5 Cruelties perpetrated by U.S. troops probably refers to the crimes perpetrated by Union forces in western Virginia, which President Lincoln would later illegally admit as a separate state in the Union.

Saturday, June 8th

We were aroused at 4 o'clock this morning by the drum. After a hurried breakfast, we took up the line of march in the direction of Shepherdstown. The weather was delightful. We arrived at Shepherdstown at 1 P.M. and was quartered in a school house. At 4 P.M. I was ordered on Picket at Butlers Mills, two and a half miles east of town at a ford, crossing the river opposite Sharpsburg.[6] The men were highly pleased with the place. (Distance 26 miles)

Sunday, June 9th

Being distant from a place of worship and confined on duty, we were deprived of the benefits of a sermon today, so the men were busy in washing their clothes.

Monday, June 10th

The boys are busy fishing, swimming and running foot races. I have never seen the men enjoy themselves better than they do here. We have seen nothing of the enemy since we have been here. Under the direction of Noble T. Johnston the boys commenced throwing up fortifications on this side of the river.

The ladies came down to see the company drill. It justly having the reputation of being the finest drilled company in the Army of the Shenandoah.

---

6 Sharpsburg. Maryland town about 15 miles north of Harper's Ferry, just across the Potomac River. Site of future major battle.

Tuesday, June 11th

Learning that a desperado by the name of Cook had organized a band of robbers, who were committing depredations upon the citizens along the border, we commenced to make arrangements to cross and surprise him, he then being at Sharpsburg. A boat had been sent to me by Captain Doyle for that purpose. Everything was ready for the adventure. Tomorrow night was the time fixed. The party was to consist of W. A. Young, N. T. Johnston, Sergt. Brown and Capt. Taylor. The men have nearly finished their earth works which will command the ford. The position is a fine one and if attacked, we could successfully resist trouble with our number.

Wednesday, June 12th

After completing all the preparations necessary for our trip tonight, we strolled along the bank fishing. At eleven o'clock I received orders to report to Col. Harper at Shepperdstown with my company. We hurriedly loaded the wagon and were soon on our way. The men had no time to prepare their dinner and the citizens took them to their houses. At 4 P.M. we moved in the direction of Charlestown, reaching it about 10 at night. We encamped in the fair ground for the night. The distance was 12 miles. (Distance 38)

Thursday, June 13th

We took up the line of march for Winchester 22 miles south of Charlestown,[7] having nothing to eat since dinner yesterday, the men began to be hungry. Col. Deavenport escorted us to breakfast. We reached Winchester at sunset and encamped in the fairgrounds. My company, being raised in Va., Col. Harper permitted us to go home for the night. General Johnston evacuated Harpers Ferry this morning, falling back to Bunkers Hill. He blew up the bridge crossing the Patomac at that point. (Distance 60 miles)

---

7 Charlestown. Site of the hanging of John Brown in 1859.

Friday, June 14th

General Johnston[8] made Winchester his headquarters. We were permitted to quarter in the town. All remained quiet. This being the first time I had been home since I came to the Ferry, I made application to Genl. Johnston to permit my company to remain on duty in this place in order that I could recruit.

Saturday, June 15th

Remained in town. Nothing of importance occurring. Citizens are censoring our leader for evacuating Harpers Ferry, but few of them sustained him. Little do they know what is necessary in military movements.

Sunday, June 16th

I took the company to church today. At 1 o'clock we received orders to report to the regiment. At 4 o'clock we marched in direction of Bunkers Hill 12 miles from Winchester, encamping on Stephenson's Farm, 5 miles north of Winchester. It rained during the night. Having no tents, we were pretty well drenched. (Distance 65 miles)

---

8 General Johnston. Joseph Eggleston Johnston. Born 1807, Farmville, Va. Graduated West Point, 1825. Spent 8 years fighting Indians in the West. June, 1860, appointed Quartermaster General and Brigadier General of the U.S. Army. Resigned from U.S. Army April, 1861 and appointed Major General of Virginia Volunteers. Assigned to Harper's Ferry, slipped away from a superior Union force. Reinforced General Beauregard at First Manassas and promoted to full general, July, 1861. Fought against McClellan in Peninsula Campaign, wounded twice at Seven Pines. Served in the West, commanded Army of the Tennessee, but was too cautious for President Davis and relieved of command July 1864. Replaced by General John B. Hood. Following Hood's difficulties in Franklin and Nashville campaigns, General Lee placed Johnston back in command and led Army of the Tennessee in the Carolinas campaign, surrendering April 26, 1865. Died 1891.

Monday, June 17th

An early hour found us again on the march. When we reached Bunkers Hill, the whole army was in line of battle. We were thrown in front. After loading our guns, we staked guns. Expecting at any moment the enemy would dash upon us, we remained here for two hours, when we were ordered to fall back. Having nothing to eat today, we were hungry and tired.

Had it not been for the generosity of the West Augusta Guards, who opened their haversacks to us, we could have been in the backgrounds. Encamping that night at Mrs. Carter's, 5 miles from Winchester, we marched 15 miles today. (81 miles)

Tuesday, June 18th

Remaining here until late in the evening, we marched to the fair grounds near Winchester. We were again permitted to go to our homes and return by tattoo. Distance 4 miles. (85 miles)

Wednesday, June 19th

Received orders from headquarters Army of the Shenandoah to relieve Captain Avis[9] Company then on duty in the town. I quartered the company in the building near the Farmers Bank. The continentals looked fine this morning as they moved off a hundred strong. Our duties in Winchester was to police the town, guard all public property and hospitals, which kept us busy. Of course, the men were satisfied with being home. The brigade moved late in the evening to Martinsburg.

---

9 Captain Avis. John Avis, b. 1818, served as 1Lt, Co. K, 1st Regiment, Va. Volunteers in Mexican War. Deputy Sheriff at Charlestown during John Brown crisis, 1859. Appointed Capt., 1st Regiment, Va. Volunteers, April, 1861. Capt., 5th Regiment Va. Inf., May 1861. Resigned June, 1862. Appointed Provost Marshal at Staunton, Va., June, 1862. Died 1883.

Thursday, June 20th

I was ordered to relieve lt. Bush of the continentals then at Strasburg on duty. I sent Lt. Fletcher and 3 men to relieve him. The duties at Strasburg were similar to those at this place. I was also ordered to relieve a squad on duty at Opequary Bridge. I sent there Lt. Jones and Sergt. Funk.[10]

Friday, June 21

Captain Silbert's Company was ordered to assist me in the duties which were very heavy at this time, the town being full of straggling soldiers and whisky plenty. Consequently, there was some riots during the night.

Saturday, June 22

Ordered by Genl. J. E. Johnston to send to Charlestown after some men who were trying to get to the enemy. Lt. Maskle was sent in charge of the squad. He only succeeded in arresting Wm. Anders, who was released by taking the Oath of Allegiance.

Sunday, June 23

Having a number of citizens in the guard house who had been arrested in Jefferson County charged with treason, I was ordered to send them to Charlestown and thare to be handed over to civil authorities. Sergt. Kuntz took charge of them.

Monday, June 24

Several new companies arrived today. I mustered them into service and formed a battalion. One of the companies was from Page [County], the other from Shenandoah [County].

---

10 Sgt. Funk. Jefferson William Obet Funk, b. 1841, brother of Colonel John Henry Stover Funk. Promoted to 2nd Lt April, 1862. Captured May, 1864. Sent to Ft. Delaware then to Morris Island, S.C., where he became one of the "Immortal 600"---Confederate prisoners placed in front of the huge Union cannon on Morris Island to protect it from fire from the Charleston batteries. Imprisoned Ft. Pulaski, December, 1864, returned to Ft. Delaware, March, 1865, where he died of dysentery, May, 1865.

Tuesday, June 25

Recruits are received very rapid. My company now numbers 84 men. When I came it numbered 45. Quite an increase. The militia has been ordered out and in compliance with the request of Brig. Genl. [James H.] Carson, I have been drilling the officers. They are a dull set of fellows.

Wednesday, June 26th

I was ordered to seize all the whisky I could find. I emptied some three hundred gallons today for different individuals. They will never forget me for it. Their curses will follow me as long I fear as they exist. The guard house and jail is crowded with drunken soldiers who have been committing depredations.

JULY 1861

Monday 1st

Nothing worthy of note has occurred in the last four days except the same routine of duty, arresting straying soldiers, quelling riots, emptying whisky and impressing negroes for the hospitals. The sick in Winchester numbers 3,000. The army is encamped around this place. The 1st brigade is encamped near Martinsburg employed in tearing up the Balt. and Ohio R.R. and destroying the cars.

Tuesday, 2

Our regiment had quite a brisk skirmish with Pattersons advance. He crossed the river early this morning, surprising our pickets and advancing in direction Haynsville. Genl. Jackson ordered the 5th to advance. They advanced as skirmishers for a few hundred yards, when the enemy suddenly fell upon them. They were engaged for nearly two hours, behaving very gallently, opposing nearly 6 full regiments of the enemy. The enemy began to take advantage of the number by beginning to out flank them.

Being compelled to retreat, they fell back some three miles again forming a line of battle. After remaining here for some time, General Jackson fell back to Darksville[1] where Genl. Johnston advanced with the army. This is the first engagement for any of the Shenandoah and I am proud that the Old 5th has reflected credit upon the Patriotic band. The Morgan Continentals under the command of Captain John Avis behaved very gallant. Our loss was killed 2, wounded 10, and missing 6. While the enemy sustained a loss of killed, wounded and missing 300. The excitement that prevailed among the citizens and soldiers is beyond description, being the first hostile engagement in

---

1 Darksville. Located about 15 miles from Winchester, 2½ miles south of Martinsburg.

this portion of the state. While those who had participated in it were busy in informing his friends of the narrow escape he had made and the heroic deeds he had performed. Half the town would turn out to see a wounded man when brought to hospital.

Wednesday, 3rd

From all I could learn, I thought an engagement would be inevitable today. Armed with a large saber and burning with a desire to see the enemy (only to be cooled in future). I started at early dawn for the army leaving Lt. Maskett in command. I reached Darksville at 12, found our army in line of battle on a ridge just beyond and was informed that the enemy was at Big Spring, 4 miles beyond.

The men were in high spirits, anxious for an engagement. Highly delighted with the affair of the day before. Many of them gave me the most minute details of the affair and the important part he performed, imagining themselves already heroes. Nearly everyone was busy penning the details of the enagement to their families and friends. I saw several prisoners captured by our force, which excited a great deal of curiosity. Late in the evening, I returned home. I found Winchester still in a state of excitement.

Thursday, 4th

No news of importance from the front, all quiet along the lines. Considered by the military men as a successful skirmish. Rumors that Genl. Patterson has fallen back. The army is still at Darksville, 15 miles from Winchester. The militia are still drilling at this place.

Friday, 5th

Not hearing from the men at Strasburg, I visited them but found everything all right. The boys by their attention to duty had won the praise of the citizens. Lt. Fletcher had enaugerated a strict discipline.

Saturday, 6th

Returned to Winchester early this morning. The 33rd regiment formed under the command of Arthur Cunnings was attached to our brigade. This regiment is formed chiefly of Shenandoah, Rockingham and Page companies, which make a very fine regiment.

Sunday, 7th

All quiet along the lines. Peter Kuntz sisters and myself visited the camps of the Alabamais, then in Baldwins Woods. The girls took with them a basket filled with presents for the soldiers.

Monday, 9th

The army fell back upon winchester encamping a mile from the town. Easley Brigade at Woods Farm. 1st Brigade on Bakers. Genl. Barta[2] and Col ___ Brigade were encamped on Baldwin's Farm. Several regiments came in today. The troops are constantly drilled.

Tuesday, 10th

Great excitement prevailed throughout the town, it being understood that Genl. Patterson had advanced as far as the yellor house within three miles of this place. Men, women and children were running to and fro throughout the town, expecting at every moment the place would be shelled.

Our troops were drawn up before the half finished earth works on the north side of town. Like the rest, I made every preparation, wore a large saber through the day and went to bid farewell to my friends. By bedtime, everything had become calmed, so we all retired with our usual quietness to dream of wars and their horror.

---

2 Genl. Barta. Probably refers to General Francis Bartow.

Wednesday, 11th

The troops are busy completing the fortification, the works chiefly performed by militia under charge of Major Whiteing, chief engineer.[3] A North Carolina regiment armied under command of Col. Anderson.[4]

Heard nothing of the enemy today. Dry goods is becoming to be a scarce arricle and at very high prices.

Thursday, 12th

Removed our quarters to a building opposite Wm. Russell's store. The building heretofore occupied has been converted into an armory for the repair of arms. We soon made it a very comfortable quarters.

Friday, 13th

The same scene enacted as on the 10th, the town wild with excitement. The enemy is again reported to be advancing. Some of the citizens have took safety in flight. By night it became quiet.

Saturday, 14th

All quiet on the front. Patterson has turned his colums toward Charlestown. One of the Kentucky regiments killed one of his company today at Philip Hoovers.

Sunday, 15th

Visited the fortifications, the first I had ever seen and we imagined that none of the enemy could scale it.

---

3 Major Whiteing. William Henry Chase Whiting, was born on an army post in the South but was raised in and appointed to West Point from Massachusetts. Graduated first in the West Point class of 1845. Tutored Stonewall Jackson at West Point. Brilliant engineer. Having married into a North Carolina family, he resigned as U.S. army major and joined the Confederacy. He held various Confederate commands and rose to Major General. He was wounded and captured at the fall of Fort Fisher in 1865 and died a prisoner in New York.

4 Col. Anderson. George Burgwyn Anderson, commissioned Colonel of the 4th North Carolina Infantry in April, 1861. Killed at Sharpsburg.

Monday. 15th [sic]

There was some commotion at headquarters which we could not fathom. A Georgia regiment arrived this morning and one battery of artillery. Tonight I met Genl. Johnston's staff going down the railroad, part of them were disguised.

Thursday, 18th

I was awakened this morning at daylight by a courier, commanded me to the General's headquarters. When I got there, he was yet in bed. His A. A. Genl. Kirby Smith[5] ordered me to cook three days rations to take one blanket and be ready to march by one o'clock. I returned, gave the order issued guns and filled cartridge boxes. We expected not to go very far from town. The enemy had orders to close their front. They had pulled down the fences for miles. This made us believe we would engage the enemy shortly. About two o'clock P.M. the army commenced to pass through the town in direction of Millwood. This led us to believe that we were going to flank the enemy by the Shenandoah River. I ordered the men to take one days rations in their haversacks and the other to follow in a wagon. At 4 P.M. I found that the regiment was several miles ahead of me, so I started with 60 men rank and file. We marched very rapidly to overtake the regiment. Some of them gave out. I used every means in trying to keep them up, but it was to no avail. I never was as much worried in my life in spite of my efforts, I could not prevent some of them from falling behind. I entreated and used every exertion, but of no avail. It was dark when we passed through Millwood, but on we marched, the men nearly all looking down. At midnight, I found that we could go no farther and encamped on this side of the river.[6]

We were uneducated to marching and it was very severe upon us. Lt. Maskett fell behind and returned to Winchester. This had some effect upon the med to see their officer behave like this. Distance marched today 19 miles. (Whole 104 miles)

---

5 Genl. Kirby Smith. Edmund Kirby Smith. Brigade commander under General Joseph E. Johnston, wounded at First Manassas.

6 The river. Refers to the Shenandoah River.

Friday, 19th

Arose early this morning and crossed the river in a boat. Genl. Kirby Smith's brigade forded the river. The men were considerable worsted with yesterday's march, but we toiled on and on until we reached Piedmont, a station on the Manassas Gap Railroad in Louden Co[unty]. We reached the station at 11 o'clock. The regiment was just getting on the train. About one half of my company was behind. I left Sergt. Grim to collect and bring the rest down on the other trains. We were very hungry, having nothing to eat since the day before. We were soon on our way. All along the road, ladies had assembled to cheer us on with their smiles and prayers, which animated the troops. This portion of our state is noted for the beauty of its women. At every station large tables were spread with provisions for us. Those of us who had nothing to eat, did it justice. Some of the scenes along the way were very effecting. It will not be forgotten by those who witnessed them.

I saw an old lady on her knees praying aloud for our success. The interest expressed in our behalf by the ladies made the soldiers hearts swell with gratitude. We reached the [railroad] junction at 4 o'clock. I did not like this place. A number of troops were encamped around the junction. Several huge fortifications had been thrown up in the neighborhood. When formed, we moved off to a small piece of the woods, about half a mile from the railroad, drew mess, pork and crackers for the night. Having no utinsels, the men roasted their meat, upon the ends of their ramrods. We were tired and exhausted, throwing ourselves upon the ground for the night and soon fell asleep. At 10, we were aroused and ordered to Bulls Run. The men could hardly walk. Several times we halted along the road. As soon as settled, we were asleep on the road. I have never enjoyed as sweet a nap as I did for a few moments on the roadside. We reached encampment at midnight. My company was detailed as picket. Distance 51 miles. (Whole 155 miles)

Saturday, 20th

We unloaded a large amount of crackers and meat. We hauled water for the brigade under the personal superintendence of General Jackson. 3 this morning, the pickets had a skirmish. At 3 P.M. the brigade moved about a mile to the right in a strip of woods. We remained in charge of a commissary store. We could see the enemy about a mile off on our front.

Sunday, 21st

This day will be remembered by all especially those who have participated in this engagement. Before we had gotten our breakfast, an order came for us to rejoin our regiment. In a few moments the enemy opened their artillery. As we started, General Jackson ordered us back. In a few moments the brigade passed us. I was desirous of participating in the engagement should thare be one. I submitted my wish to the men who at once consented so off we went without orders. After several hours march through high grass, swamps and woods, we found the regiment in line of battle. Once I came nigh marching into the enemies lines, but I discovered my mistake while I had yet time to return. The cannon fire had now become terrific. Now and then I could hear rattling of small arms, seeming to be nearer and nearer. At 9 our brigade was ordered forward. The day was hot and we advanced with some rapidity. After advancing a mile or more, a shell passed over our heads, being the first I ever heard. I need not say that the sound was unpleasant to an inexperienced ear. Wounded men were carried back, which is never very encouraging to troops marching to the fields. They told conflicting statements. Camtons [sic] Legion had engaged them and been whipped. We were immediately ordered to their support. The troops were then run through our lines in every direction. Our regiment was on the right. Next the 33rd, 4th, 27th, and 2nd. We were ordered to lay behind a hill just in front of the advancing foe. The bullets flew over us like hail. While the ear was deafened with the sound of hundreds of shells bursting in the air and throwing the fragments whisteling through the air in every direction, giving things in general a dangerous appearance.

We lay in this unpleasant position some twenty minutes, some praying and some swearing. Then we were ordered to advance, which we did under a murderous fire of musketry shell and grape. When we reached the top of the hill the enemy were discovered advancing in large force. A regiment of Bivouics [Zouaves] were trying to flank us in the left. I ordered my men to fire. They at once commenced to fall back. Upon our front was a Michigan regiment, bearing a secession flag and upon our approach they gave us the signal (which was raising the hand to the head and repeating "Our homes"). They appeared to be falling back with their backs to us. We at once took them to be our friend. They turned around and gave us a volley. Fortunately for us, they aimed too high and their balls passed over our heads. We drove them back several hundred yards. They were then reinforced by several other regiments. Col. Harper, finding our position rather precarious, ordered us to fall back to a strip of woods. I did not hear the order until the men were half way across the field and came nigh being taken prisoner, but when I did discover it, my conscious, I left in a hurry. I was afraid I would be shot in the back. Turning around, I thought I would try to back out of it, but this was not fast enough. I turned around and the way I did scamper was a caution. We soon reformed and advanced again, but some of the men had not stopped falling back yet. On we went and was forced to fall back again. We again formed in rear of a thicket some reinforcements had arrived. General Beaureguard[7] came riding up, ordering us to advance.

Leading the head of the column in person while climbing a fence a cannon ball came sweeping a long striking the rails beneath me scattering me and the rails both into the hollow. I at first thought I

---

7 "General Beaureguard." Pierre Gustave Toutant Beauregard. Born 1818, St. Bernard Parish, Louisiana, of Creole parents. Graduated West Point 1838. Served in the Mexican War, wounded twice, breveted twice. Appointed Superintendent of West Point in 1861. Accepted commission as brigadier general in Confederate States Army March 1, 1861. Commanded shelling of Fort Sumter at outset of the War. Commanded Confederate troops at First Manassas. Promoted to full general August, 1861. Transferred to Western theater, served under General Albert Sidney Johnston and commanded at Shiloh when Johnston was killed. Commanded coastal defenses in the Carolinas and Georgia. In April, 1864 joined General Lee in the defense of Richmond. Defeated General Butler at second Battle of Drewry's Bluff. After the War, returned to New Orleans and died 1893.

was killed but soon discovered that I was uninjured. A battery was then pouring destruction into our flank. General Beauregard ordered us to charge it. It was here. His horse was shot from under him. General Bartow[8] came riding up and when interrogated by General Beaurogard "How was the day?" he replied whispering "like hell." Then pointing to our brigade, he remarked, "Look at Jackson. He stands like a stone wall."[9] General Bartow wheeled his horse and tried to rally his troops, when he fell mortally wounded. As we neared the battery, they hurled grapes into us. Genl. Beaurogard told us to dodge them. When we were within some two hundred yards of them, they fell back. We then marched by the left flank opposite a house where Sherman's battery[10] was stationed. We charged down the hill and took it. In conjunction with the fourth regiment on our left, driving the enemy back to a lane. They began their preparation retreat. A long line could be seen moving in direction of Alexandra. We were soon in pursuit, hurling missiles of death into their retreating ranks. We followed them to the stone bridge, when we were ordered to the junction, it now being dusk.

Had the enemy have made another advance, we could not have resisted them. I then rode over the field in quest of those of my company who had fallen. Never in my life have I seen such a spectacle. The dead and wounded were strewn for miles over the field. Here and there an arm, a leg, men and horses were piled together. The wounded filled the ear with their pitying groons. I have never beheld more affecting scenes than this. To think that men in the most excruciating pain had no one to do or receive their last words.

---

8 General Bartow. Francis Stebbins Bartow. Born 1816, Savannah, Georgia. Graduated Yale Law School. Planter and slave owner. Elected to the Provisional Confederate Congress. Captain, Oglethorpe Light Infantry at seizure of Fort Pulaski. Transferred to 8th Georgia Infantry. Led his troops in a charge down Henry Hill at Battle of First Manassas and was mortally wounded and died, July 21,1861.

9 "Look at Jackson. He stands like a stone wall." Captain Funk here gives the only known eye-witness account of how Stonewall Jackson received his famous nickname. See Introduction.

10 Sherman's battery. Then Col. William T. Sherman, commanding a Union brigade with attached artillery.

It was here that I met with Captain Rickets,[11] whose leg had been broken by a ball. A surgeon was preparing to amputate it. When I interfered and told him it could be saved, I bound it up and sent him to the hospital. I found a horse tired and worn out. I mounted him and rode from wounded to wounded man binding up his wounds and consoling him as much as I could.

Some 50 of us were at the house of a lady who had been wounded 7 times. Every piece of furniture was riddeled by bullets. No one was present but her daughter who was absent during the battle. The house was between our line of bullets and those of the enemy. Just as we were about leaving, a shell bursted out side and we thought it was the enemy advancing. Those of us who had horses mounted and rode off in a hurry, while we found it to be an accident.

In administering to the wounded, I came to a Lt. by the name of Bonds. I discovered that his wound was mortal. I told him he had but a short time which to stay with us and he had better make his peace with his God. He turned upon his side and expired. I found near him a small and handsome pistal which I carried off as a trophy with the idea of presenting it to V____. I shall not attempt to pen the heart rendering scene of this cruel day. I turned in disgust and rode in direction of the junction. A mile from the field, I met President Davis with the reserve. The roads were filled with wagons and ambulances carrying off the dead and wounded. Taking one wounded man behind me, I went on in a trot to the junction. When I got there, I could find no one I knew. Thousands of stragglers crowded every place.

I hunted in vain for my Sergt. and the only one I found was Ober, a member of my company, who was crying saying his Captain was killed and he was going home. I then went over into the woods and went to sleep.

---

11 Captain Ricketts was a Union officer, as was possibly the Lt. Bonds mentioned later.

Monday 22nd, 1861

Yesterday we engaged and defeated one of the best equipped armies ever martialed, flushed with anticipated success and proud of their appearance, they rushed upon us. The drawing of our men turned them back as often as they advanced and after ten hours of hard fighting, they made a presumption retreat for Alexandria, where only a few days before they rallied forth as they thought to success, with colors waiving and bands playing their favorite martial strains. The officer had made the men believe that the few ragged rebels would fall upon their knees at the sound of "Yankee Doodle" and play homage to the stars and stripes or abandon their arms and fly before the splendor of their military appearance.

It surely was a grand sight to have seen their army move off under such circumstances, each soldier eager to see a running rebel. Their return was much less pleasing to their friends. For twenty miles a retreating army the road filled with wagons, artillery, horses, and men. Arms, amblances, cloathing and supply stores spread in abundence all along the road.

Their advance was met with more resistance than they anticipated. The ground was covered with their killed and wounded. A large number of citizens had followed the army to join in the celebration of their achievements. Great preparations had been made for these entertainments. Enormous quantities of wines and refreshments were captured by us intended for that purpose.

Many of these citizens were captured. The road running from the battlefield to Alexandria was filled with abandoned property. Ambulance drivers would cut their horses loose and make their escape.

Even in our brilliant success, we must stop to mourn the death of so many braves who have fallen upon our side. Meeting a soldiers death and filling a soldiers grave, our victory was clear, our loss being heavy.

Our regiment lost and killed, wounded and missing 64 men. My company, 1 killed and 4 wounded.

Benjamin Kaufman, a brave and gallant member of my company, was killed in our charge upon Ricketts Battery. He was pierced through the head. He was a citizen of Maryland, 28 years of age. C. H. Newton, E. Castleman and James Gem were wounded.

The Continentals suffered but little. The Winchester Rifles had 4 or 5 killed, 6 or 7 wounded. Among the killed was their Sargent, a promising young man from our place, Owen Bergess. Among the wounded was their Captain, W. L. Clarck. I arose early this morning and turned my attention to the wounded, which were coming in by the hundreds, binding up their wounds, extracting bullets and etc. It had rained all day and was quiet disagreeable.

At 10 A.M. I was ordered to rejoin my regiment, which was then ordered back to camp. I found nearly all the men had come in. The roads were very muddy. We reached camp at noon; having no tents at all, I cut up several bolts of full lensey I had received for blankets and gave them to the men to make tents out of, which made them very comfortable.

I found it impossible to procure a coffin for Kaufman. In the evening, I ordered out a detachment to bury him. They rolled him in a blanket and placed him in his long home. A large number attended his funeral rights. Many expressed their gratitude at the manner he was buried. His grave was marked so that if ever desired by his friends, he could be sent to them.

General Jackson was wounded in the middle finger of his left hand, breaking the bone. The rain had made our camp very disagreeable. The army was busy burying their dead and writing letters to their families and friends.

The whole force engaged upon our side did not exceed 12,000 men, while the enemy was supposed to have 30,000 engaged. Distance 20 miles. (Whole 175 miles)

Tuesday, 23rd, Camp Jackson

Still rainy, cleared off about noon. Still burying the dead. Ordered to discharge our pieces and clean them up. Our army was very much disorganized. Water very scarce and unfit for use.

Friday, 26th

Nothing of importance occurring. I sent an order for Lieutenant Fletcher to rejoin his regiment. Many of our friends, hearing of our engagement, had come down to see us. My family had been told that I was killed which, of course, made them very uneasy.

Saturday, 28th, Camp Johnston

We moved our camp some 4 miles to the west of our present camp called Camp Johnston. Water and wood was very scarce.

Sunday, 29th

Attended preaching at General Jackson's headquarters. Lt. Fletcher[12] and his squad rejoined us bringing their tents with them. We were visited daily by our friends. I am somewhat unwell today. (Whole miles 179)

Monday, 30th

My company was mustered for pay by Col. W. H. Harman, unfit for duty. Officers held a meeting in reference to their field officers inviting Col. Harper and Harman to resign. I declined to attend, being from another country and had nothing at stake in this issue.

---

12 Louis Johnson Fletcher, silversmith, age 18, enlisted April 18, 1861 at Winchester, 2nd Lt. 1st Lt. Feb. 1862, Capt. April 1862. Fatally wounded at Malvern Hill July 1, 1862.

AUGUST

Friday 3rd, Camp Harman, Fairfax County

We took up the line of march this morning from this camp passing by the battlefield and stone bridge. We saw where the enemy had barracaded the road. Knapsacks, wagons, boxes, guns and artillery still remained along the road to tell that this was the way the Yankees went in their retreat of the Twenty First. We passed several large hospitals filled with Yankee wounded.

The day was exceedingly hot and the roads dusty. The march was very hard upon the troops. We passed through Centerville, a village of nearly two hundred inhabitants. Here is where my mother was born. I am confident that she could not recognize the place of her infancy, for the country is white with tents as far as the eye can reach. We went into camp one mile beyond the village which was called Camp Harman, after John Harman, Chief Quarter Master of our brigade. We at once commenced clearing of the ground and pitching our tents in regular order. The camp promises to be a pleasant one. Distance 10 miles. (Whole 199 miles)

Saturday, 4th

Men are busy today in cleaning up the camp and making themselves comfortable. Billy rejoined us, bringing down some men from Winchester. My company numbered at this time about 50 men for duty. Lt. Masket arrived, he being absent without leave.

Sunday, 5th

Attended preaching today in the 27th regiment. The weather is fine. All seem to like the camp very well. The enemy occupy Arlington Heights.

Monday, 6th

Ordered out today in review before the President. I have never in my life suffered as much with heat as I have today. The 1st Va.

Regiment passed in review before him. We then returned to camp. The men are enjoying themselves finally, having sham battles, dress parades, concerts, etc.

Wednesday, 8th

Ordered to cook a days rations and be ready to move in an hour. At the appointed time we moved in the direction of Fairfax Court house. We marched through that place, marched 4 miles beyond and was thrown into a line of battle in a large field. We lay there half an hour and then was ordered back to camp. We reached camp at midnight. Many of the men were so exhausted that they could not reach camp. The alarm was false. Distance 22 miles. (Whole 221)

Saturday, 11th

The regiment was paid off by Virginia up to the 30th of June. My company was not paid because it was not mustered for July.

Sunday. 12th

Had inspection this morning. Col. Harper and Col. Harman were absent. I was left in command of the regiment. Attended church today. Several gentlemen came down from Winchester with provisions for the company.

Tuesday, 14th

The company had become greatly dissatisfied and blamed me to a considerable extent. Lt. Maskell seemed to fan the flames. He persuaded the men that they were not mustered into service and that they could not be held. This was urged by those who were anxious to leave it. Men had ran off and more threatened to follow. It seemed as if nearly every friend I had deserted me save Kuntz, Grim, Baker and a few others. Meetings were held drafting resolutions to ask me to let them return. Others were held pledging that they would return home if they were not paid by a certain time.

Not only was the fuss here but at home. Those who knew nothing about it censured me severely. Parents wrote for their children to return and insulting letters were received by me, blaming me for the whole. I used every means to satisfy the men. Some would not be convinced. I had once concluded to resign, but I thought that would only convince the public that I was guilty of something. So I became determined to defeat my enemies.

One great reason for this dissatisfaction was that they did not get payed off when the rest did. The facts were plausible. We were on detached service in Winchester when the rest were mustered for pay. Consequently, we had not the opportunity until we were at Camp Johnston. I got several letters from the pay master general, who stated that my rolls had been received and were correct. This satisfied them some, only to be aroused by Lt. Maskell again.

Sunday, 19th

We were again ordered to support the pickets. We advanced some 4 miles in direction of Fairfax Court House when we were ordered back to camp. The men were kept actively employed, drilled 4 times a day beside dress parades and police duty. Distance 8 miles. (Whole 229 miles)

Tuesday, 21st

Had a review of the regiment before Col. Harper. Men were still dissatisfied. It is reported in Winchester that they are all coming home. The thing has me worried to death. Knowing that it is not my fault, but few men would have endured what I have, but I am determined to beat my persecutors.

Thursday, 23rd

The brigade reviewed and drilled before General Joe Johnston. It looked very well the general Complimented them very highly. Our General is looking quiet well. I consider him one of the finest looking military men I ever saw and a very handsome rider.

Sunday, 26th

After our usual inspection, we attended divine service. Nothing occurring in camp worthy of notice. Dissatisfaction is becoming more alarming. Mother and father urge me to resign, but I intend to surmount this difficulty and will show to the world that I can yet bring them out straight.

Tuesday, 28th

We were again ordered to march down the road in direction of the Court House. We got within a mile of the place when we were halted for some 3 hours and then ordered to return to our camp. This was not very pleasant, eternally traveling up and down the road. The pickets had attacked Long Street [sic], but were repulsed. Distance 12 miles. (Whole 241 miles)

Thursday, 30th

Ordered on picket one mile from camp. The pay master arrived this evening to my great relief. I found my rolls correct. Now I am all right again. Some of the men are somewhat ashamed of their conduct. It has kept me very busy to make them out and having them signed.

Friday, 31st

The men received their pay from the 18th of April to the 31st of June, from the State of Virginia and above all put at rest the dissatisfaction that had come nigh destroying the organization of the company. Mustered for pay for the months of July and August.

September

Saturday 1st

Finished paying off the men, felt very unwell. Slept unsoundly. Great dissatisfaction in the regiment in regard to Col. Baylor. Nearly all the companies had petition for his restoration to the command of the unit. I declined in participation in the affair. I assigned a squad of men to Winchester to arrest deserters.

Sunday, 2nd

Unable to attend church today. I have symptoms of typhoid.[1] I attribute it to mental excitement, worried so much with the dissatisfaction of my company and trying to do my best for the men I love as brothers and at last succeeded in getting justice, has so relieved my mind from its load of anxiety that I fear it has settled into a fever. Dr. Smith has pronounced it typhoid.

Thursday, 6th - Camp Beauregard

We moved our camp this morning to Camp Beauregard 1½ miles south of Fairfax Court House. They hauled me in an ambulance. The camp promises to be a very pleasant one, chestnuts in abundance. I am decidely worse tonight than I ever have been. Compelled to keep someone in my tent with me, at times delirious. The officer and men of my company have urged me to go home, so has our surgeon and Col. Harman. I declined them. Think I will be well in a few days and do not desire leaving the boys. The men are anxious that I should do. Distance _ miles. (Whole 229 miles)

---

1 Typhoid fever is an infection caused by the bacterium salmonella typhi, usually attributed to unsanitary conditions. The bacteria infect the intestines and spread to the bloodstream, causing high fevers. Headache and constipation followed by bloody diarrhea usually results.

Friday, 7th

My fever has increased. My friends still advise me to go home. Unable to get out of bed. Col. Harman has forwarded my application. He has received his promotion as Col. Major Beauregard, Lt. Col., Absolem Koenie Major.

Wednesday, 14th

My fever increased until I became very ill. My friends became alarmed and at last I was pervailed upon to go. It was with reluctance that I went. Lt. Fletcher, the only officer left in command, was unwell himself and I fear the boys would be neglected. At 10 A.M. we arrived at Fairfax Station. Had to wait until evening for a train. Private Duncan accompanied me. I was put into a warehouse, where a number of Yankee prisoners were confined. Late in the evening we arrived at Manassas too late for the train, so I was put into an old barn fixed up for a hospital. I anxiously awaited the morrow.

Thursday, 15th

Forbid that ever I should spend such a night as I did last night. I slept but little delerious all night. Kept Mr. Duncan busy to keep me in bed. At 5 o'clock P.M. I was placed upon the train bound to Strausberg. We took dinner at the plains. At 9 P.M. we reached Strausberg. I was carried to an ambulance which was bound for Winchester. At 12 midnight we reached Winchester.

Friday, 16th

My family were surprised on my unexpected return, not hearing of my illness until I had arrived. I was at once carried to bed and a family physician called in. My friends were excluded from my room. I also found my father very ill, but gradually recovering.

OCTOBER, 1861

Tuesday, 4th

Today is the first time I have been able to leave my room, during which time I visited my friends, who a few weeks ago were condemming me severely for the dissatisfaction in my company. They had learned that it was not my fault. I had received several letters from my company. They were having an unpleasant time of it on pickett. They looked anxiously for my return.

Wednesday, 5th

I ventured out in a buggy today. Went to see the ladies Aid Society, who were making clothing for our company. We are very much indebted to our women for their kindness in furnishing such clothing to protect the soldiers from the blasts of winter.

Saturday, 6th

Nellie and I went visiting this evening to Stevensburg. I am gradually growing stronger. Expect to return to duty shortly. During my absence, the army had fallen back to Centerville, also the battle of Balls Bluff[1] has been fought, we gaining a decided victory. General Evans was in command. Articles of clothing are beginning to be scarce and high.

---

1 Battle of Ball's Bluff. Confederate forces under General Nathan "Shanks" Evans defeated an attempt by Union forces to cross the Potomac River at Harrison's Island and capture Leesburg.

Thursday, 13th

My furlough expired today, but the surgeon pronounced me unfit for duty and had it extended.

Was made a Master Mason in Hiriam Lodge this evening. Ordered Lt. Maskell to rejoin his company, he being sent off on sick furlough and now well. He has overstayed his time and been behaving badly.

Monday, 24th

Left Winchester to rejoin my regiment. Weather very cool. Arrived at Strausberg at 10 P.M. Stopped at Richards Hotel.

Tuesday, 25th

Aroused very early. Very cool and a heavy frost. The train left at 2 A.M. Breakfasted at the plains [sic]. Reached the junction at 9 A.M. Found Sargent Baker in charge of extra baggage at the depot. Gov. Letcher,[2] Z. Kidwell and William P. Thompson accompanied him. In the evening, we went over in an ambulance. Found the boys in good health and spirits. They had a custom of placing those upon a barrell who arrived with new clothes. Sergt. Pritchard had to pay the penalty for one of the most ragged men in the company escorted him around the brigade. Upon my arrival with a new suit the idea was suggested. The regiment had marched during my absence 33 miles. (Whole 302 miles)

Wednesday, 26th

I found the company in very good condition. The men were very anxious to get home. Lt. Fletcher has gotten along remarkably well with them. They had forgotten the dissatisfaction that had existed a few months before.

2 Governor John Letcher of Virginia

Thursday, 27th

All the Virginia troops in the army of the Potomac were marched to the fortifications where they were addressed by Gov. Letcher, who presented unto each regiment a handsome Virginia flag. Among the souls present was Johnston, Beauregard, [William] Smith, Jackson and Longstreet.[3] After the flags were presented each regiment drilled before the Governor and Generals. About sundown we were ordered back to our camps.

Friday, 28th

We resumed drill and practiced review for on the next day we were to review before the Governor.

Saturday, 29th

This afternoon we were marched to a field beyond Centreville and formed in line. General Longstreet was in command. Our General Jackson was present riding his old sorrell and mostly by himself without an aide. While a cavalry company followed the rest. This caused the men to admire Jackson. After the line was formed, the Governor in company with General Johnston and staff, rode along our front and rear. Then we moved of by the right flank and passed in review and then returned to camp.

Monday, 31st

We were today mustered for pay for the month of September and October by Major Kolnir, mustering officer. The men are very anxious to go with General Jackson to the valley. He has been promoted to a Major General and ordered to command the Valley Dept.

---

3 Longstreet. General James Longstreet, b. SC 1821. Graduated West point 1842. Wounded at Chapultepec during the Mexican War, brevetted twice for gallantry. Confederate brigadier general, June 17, 1861; major general, October 7, 1861. Around this time was promoted to Lt. General and commanded a corps of the Army of Northern Virginia through the war. Gen. William Smith was a former and future governor of Virginia.

November 1861

Tuesday 1st.

The regiment received two months pay, for the months of July and August, which will make card playing livelier. A practice to which the soldier is very much addicted.

Met Gov. Letcher at Col. Harman's Headquarters.

Wednesday, 2nd

Ordered on picket at Oxfords Road. Left camp early this morning and relieved the fourth regiment then on picket. Col. Harman sent me on the outpost with his companies. It commenced raining and blowing as soon as I posted the men. I never saw such a storm. It blew down all the tents in the encampment. It was very cold and the men were not allowed any fire. The enemy were just on our front. Distance 7 miles. (Whole 309)

Thursday 3rd

Relieved this morning by Captain Avis. It had rained all night. I found it as disagreeable at the reserve as it was in the out post. Col. Harman and myself visited Genl. J. E. B. Stewarts[1] Hdquarters. I found the General to be an exceedingly plesant gentleman. His lady was with him.

---

1 J.E.B. Stuart, cavalry commander of the Army of Northern Virginia until mortally wounded at the Yellow Tavern, May 11, 1864.

Friday 4th

General Jackson bid us adeau today. His remarks were few, but very effective. He has gone to take charge of the Valley Department. The boys are very much disappointed by not going with him. Some were very angry about it. I still believed and had hoped we would soon be ordered over.

Saturday 5th

Sent on picket again. The weather but little better. A continued firing of skirmishers today.

Sunday 6

Relieved by the 27th Va. Infantry. Returned to camp. Found that the storm had blown down every tent. The men were very angry about not being ordered with General Jackson. (316 miles)

Tuesday 8th

Captain Newton and my company were again sent on picket 3 miles from our encampment. Being unwell, I remained in camp.

Thursday 9th

Were relieved. While in Col. Harman's tent this evening, I heard him say we would start to the valley in a few days. I communicated the fact to the company, which sent them wild with joy at the news. Many of them would not believe me.

Wednesday 10th

Busy in making preparations for the march. The men were in high spirits, anxious for the order to move. Col. Preston of the 4th Va. Infantry in command.

Thursday 11th

Could not move for want of transportation. Genl. Johnston very desirous that we shall not go. Col. Allen was ordered to take charge of 2nd and Fifth and proceed tomorrow to Winchester and Col. Preston would follow with the remainder of the brigade.

Friday 12th

Moved this morning at 11 A.M. in direction of Manassas. The roads were very bad, almost impassable, from the rain we had a few days before. Arrived at the station at 6 P.M. After getting our rations, we mounted the train with our baggage. The night was very cold and disagreeable. Distance 40 miles. (Whole 356 miles)

Saturday 13th

Arrived at Strasburg this morn. at 4 o'clock. Raining very hard. One of the members of company (F) fell from the train and was killed. After loading our luggage in wagons, we took up from the line of march for Winchester. The rain continued to fall and the roads were very disagreeable. At Middletown, we took dinner. Many of the men became intoxicated. Arrived at Kernstown late in the evening. A number of men went ahead into town. The General ordered us to camp this side of the town. The men, wet and cold, pushed ahead through the pickets and would not obey orders until several men were arrested. Nearly all the men had flanked into town. I had but one man with me. I told him to go to some place in the neighborhood and I got permission to go home. The camps were very disagreeable and wet. Distance 40 miles. (Whole 396 miles)

Sunday 14th, Camp Kernstown, Frederick County

Early this morning I returned to camp, but few of my men had returned. Officers who had went to town without permission. As soon as I reported that I had arrested them, I was ordered to relieve them. We commenced cleaning up our camp and pitching tents. Our men are well pleased with the locality. Many of our lady friends visited us. Lt. Maskell was placed in arrest. I attended church in Winchester.

Monday 15th

Commenced drilling 4th, 33rd and 27th regiments came up. The encampment is a very pleasant one with the exception of wood, which is very scarce.

Friday 19th, Camp Stephenson[2] near Winchester

Remained in Camp Kernstown until today. Drilling and performing other duties. When we moved to Camp Stephenson 4 miles beyond Winchester on the Martinsburg Campsite. We had anticipated a trip to Romney. The Yankees being there under command of Kelly. Our present camp is a beautiful place and in a short while we had our tents pitched. Distance _ miles (404)

Saturday 20th

General Garnette[3] arrived today and took command of our brigade. The officers called upon him. I found him a very nice and agreeable gentleman. He is a Virginian, serving in the old army, but since has been on duty in the Penensula. A man about 26 years of age. Captain Wingate is his adjutant Genl. and Lt. Williams is his aide.

Sunday 21

Attended church in Winchester. The weather is very pleasant. The enemy is still at Romney burning property and pillaging the country.

---

2 Camp Stephenson. Farm of James W. Stephenson, 4 miles north of Winchester.

3 General Garnette. Brig. Gen. Richard B. Garnett, born VA 1817; West Point, 1841; western U.S. Army service; took over command of the Stonewall Brigade as indicated. Arrested by Stonewall Jackson after retreating at Kernstown without permission. Assigned to another command and commanded a brigade at Gettysburg in "Pickett's charge," when he was killed in action.

Monday 22nd

Our new General inspected the brigade today. The men have expressed their appreciation for him. Began to snow this evening.

Tuesday 23rd

The brigade was reviewed today by Genl. Garnette and staff. The old brigade looked very well today. A large number of citizens were present at the review.

Monday 29th

Our soldiers had committed some depradations upon the citizens when under the influence of whiskey, which was sold in abundance near the brigade. General Garnette ordered me and Lt. Lewis to destroy all we could find so we took a squad of men and went to a still house some two miles below us upon the railroad and destroyed several barrels.

We found a quanity in the neighborhood which met the same fate. The citizens will never forgive me for it.

December 1861

Wed. 1st

Four companies from our regiment were sent down to Potomac under charge of Major Paxtin[1] of the 27th Va. Regiment. They took the train at Stephenson's Depot.

Monday 6th

The companies which were detached have returned to the regiment. This camp is quiet a pleasant one. Our friends visit us often and bring with them dainties from home. Besides we visit Mrs. Carters and pass our evenings off very pleasant.

Wednesday 8th

Companies D, L, and C have been assigned on picket. The weather is cold and dry. The enemy are on the opposite side of the river.

Saturday 11th

The brigade was inspected and its deficiency in arms and accompement were supplied. Our orders were rigid. We were required to keep the men in camp. It was with great difficulty that I could keep the men out of my company in camp. They being from Winchester and close to their homes and sweethearts.

---

1 Elijah F. Paxton, who later became brig. Gen. and commander of the Stonewall brigade. Killed at Chancellorsville.

Monday 13th

Received orders to be ready to move tomorrow at daylight. No one knew in what direction. One imagined this one, some that direction. So rumors of all shapes and size that the mind could conceive were afloat. It was a late hour when the men had finished their cooking.

Tuesday 14th

In compliance with orders issued last night, we were up ready to move the march in direction of Martinsburg. Our men laying idle in camp had not been accustomed to marching and it was with some difficulty they reached camp. We encamped at the "Big Spring" 4 miles from the town, with orders to move tomorrow. (Distance 422)

Wednesday 15th

Left camp early this morning passing through Martinsburg and taking the Williamsport road. We passed the place where the Hainesville fight took place on the 2nd of last July. Turning to the left 8 miles from Martinsburg on the road leading to Dam No. 5. We lost the road and continued to march until late at night, when we were halted in a woods near the dam. We were not allowed to build fires. Distance 22 miles. (444 miles)

Thursday 16th

At daylight we were ordered a mile to the rear where we cooked. The night was very cold, being in tents all winter, it was very hard upon the men. A working party was sent down to work upon the dam. The enemy occupied the heights on the opposite side. A number of the men would fire at them across the river. We remained here until after dark. When unobserved by the enemy, we returned to the place we had occupied that night previous. The weather had become much colder, but we were allowed fires. Adjt. Bumgardner and a number of us visited the dam, where the working party waist deep in water were undermining the dam. We could see the enemies camp fires on the opposite bank. During the day, they had exchanged shots. Our men were persecuted by a large mill.

Friday 17th

At daylight we returned where we had been the day before. The weather still cold. A brisk firing kept up on both sides today.

The militia under the command of General [James H.] Carson were five miles below us opposite Williamsport. About two heavy firings were heard in that direction. We could see the shell bursting in the air. It lasted only for a few hours and all became quiet. After dark, we again returned to where we lay the night before.

Saturday 18th

Returned and cooked rations. The General [Garnett] desired us to have volunteers to work at the dam. I took charge of the party which was to go tonight. The regiment was sent on picket. The enemy had by this time discovered our object and sent reinforcements. They were discovered this morning planting a battery. They soon opened up on the regiment, doing but little damage. Higher up the river was a barn on the opposite side occupied by the enemy. The General, late in the evening, moved around with a section of the artillery. Fired upon it and soon disposed of them. A mile farther up the river was a village called Little Georgetown. I learned that some of the lones [sic] visited it during the night, being very anxious to capture them, I communicated the fact to some of my men who were as anxious for the adventure as myself. I made application to the General, who approved it, and sent it to General Jackson, who granted the request. Late in the evening the enemy concentrated their fire upon the mill, driving the working party away. The mill was soon in flames. We were at once ordered in that direction, which fusterated my plans. We lay in the same place as before. The weather had become entirely cold.

Another working party was sent down to complete the work.

Sunday 19th

At 4 this morning the work had succeeded and the dam was reduced. A fire had been prepared for those who had been imployed in the work. It was so cold that their clothes was frozen stiff before they could reach the fire, which was but a few hundred yards from the river bank.

The Hibermians, a company in the C 7th Regiment composed chiefly of Irish and commanded by Captain Robertson had been imployed upon this duty. When they got to the fire, whiskey was issued to them and the General gave them 7 days furlough to go where they pleased. At daylight, having finished our duty to the satisfaction of our General, we started back to our camps, passing through Martinsburg where our friends received us with cheers, but our enemy gave us many a black look.

This place being composed chiefly of hands who depended alone for support from the railroad and since it had been destroyed, they had fled to the enemy, leaving their families behind. It was those who looked upon us with contempt. After arriving at camp, I returned to town to attend lodge where Col. Ashley was to take a degree. I found many of the citizens very indignant. Returned to camp late. Distance 22. (Whole distance 466)

Monday 20th

Marched to our camp near Winchester. Distance 18 miles. (Whole 484)

Thursday 24th

Tomorrow being Christmas and the men are desiring to spend their Christmas with their families and friends, I made application to Genl. Jackson to permit us to go to our homes in Winchester, which was granted. The men are pretty lively in camp today. Eggnog and whiskey punches are pretty plenty.

Friday 25th

Early this morning we started for Winchester. I remained at home to dinner in order that the whole family might be collected once more, it being the first time for years. It may be the last time, specifically in times of trouble like these.

Quiet a lively day. The town is thronged with soldiers and with but little disorder.

I visited my friends which I had been accustomed to visit in times of boyhood. The scenes recalled the pleasant scenes of the past and caused me to think of one year ago.

Saturday 26th

Remained in town all day. Nothing occurring. Weather fair.

Sunday 27th

Returned to camp. It was predicted that we would be ordered to Mannassas again. Great preparations in camps every artical needed that can be furnished is supplied.

Monday 30th

Taken sick with jaundice and sent home. Preparations still going on in camp.

Tuesday 31st

Last day of the year. Late this evening, I learned the troops were ordered to be ready to move at daylight. Being unwell, yet I wished to be with my company, so I reported for duty. Captain Carpenter and I went to take leave of Mrs. Carter, where we had spent many pleasant evenings. We knew not where we were going, but various rumors filled the camp. The men were up until a late hour preparing for the morrow. This ends the company of 161.

Col. J. H. S. FUNK.
5th Va. Inf.
Stonewall Brigade
Son of
Christopher & Eliza Funk
Born June 25, 1837.
Fell mortally wounded
Sept. 19, 1864.
Died Sept. 21, 1864.

Lieut. J. W. O. Funk.
Co. A. 5th Va. Inf.
Stonewall Brigade
Son of
Christopher & Eliza Funk
Born June 30, 1841
Died at Fort Delaware
May 17, 1865.

Campaign of 1862, January

Wednesday 1st

In compliance with orders issued yesterday, we were up and had our wagons loaded ready to move by daylight. The sun arose clear this morning and just as she arose above the horizon, we were on the move taking the pike in the direction to Winchester. None knew whether we turned to the right at Mrs. Carters and striking the Pew Town Road 6 miles from Winchester, going in the direction of Bath.

It was a beautiful day and with many romantic scenes along the road, made the march very interesting. We encamped three miles beyond Pen's Town. Distance 20 miles. The troops were very much worried. Our wagons were late coming up and the night very cold. Whole distance 504 miles.

Friday 2nd

A skift of snow covers the ground this morning. The weather is very cold. We left camp early taking a road leading to Bath. It was very disagreeable marching over a miserable road, crossing unbridged rivers. It seems almost impossible for our wagons to pass over this road. We bivowaced at Ungers Store 12 miles from Bath. Distance marched 18 miles. The weather has become intensely cold. Gillims and Lolifens Brigades passed our camp. It was late when our wagons reached us and the men suffer much from the cold. Whole distance (522 miles)

Saturday 3rd

We were ready to move early this morning but had to wait until the other troops passed us. The militia under Genl. Carson[1] from Martinsburg joined us. Genl. Laney's command passed to the front about nine. We moved oft, but were compelled to halt every 20 or 30 minutes to permit the wagons to go ahead. The road was lined with broken down teams. Tent and baggage were burnt by the quantities. It was so cold that at the halts the men were compelled to make fires. The fencing for miles was burnt. In fact, it was a string of fire all along the road. It was after dark when we reached camp. Distance 3 miles. It commenced snowing at sunset and continued until late in the night. The cold had increased and now it became very disagreeable. I had just gone to bed when I was ordered to report to the Genl. F. O. Day [sic] and had to post the picket at midnight and relieve them at daylight. Whole distance 530 miles.

Sunday 4th

It had ceased snowing this morning, but the cold had increased. I relieved my pickets early. We crossed the river on a rail bridge. The road was very slippery. We were halted and ordered to tie a white strip of cotton on our left arm as a badge. I procured a sheet from a farm house for that purpose. Our march was slow, tedious and disagreeable and was compelled to halt often. Our advance guard had a skirmish yesterday, killing several and taking a number of prisoners.

When within 4 miles of Bath, we were halted for some time in order that the troops in front could be placed in position. One brigade was sent on top of the mountains to try to cut of their retreat, late in the evening, but our forces made a decent upon the place, but the enemy had retired leaving us but a small rear guard, part of which was captured. I had went ahead of our troops into the town. We captured a large amount of commissary stores. Our Brigade was then marched into the town. Our wagons had not come up on account of the bad roads. We were quartered in houses opposite the hotel. The ground was covered with about six inches of snow.

---

1 James Harvey Carson, general of VA militia

Fortunately for us, the houses we occupied was the headquarters of the Yankees and in their flight they had left some commissary stores and we subsisted upon them. Until our wagons came up. The large hotel was the property of Mr. Strather[2] alias Port Grain, who was serving in the Yankee Army. In his absence our men pillaged the house and destroying a great deal of the property before our officers could prevent it. Distance 4 miles. (Whole distance 534 miles)

Monday 5th

Early this morning we marched back to our wagons a mile from the town where we cooked our breakfast and a days rations. The wagons were repacked and we started in pursuit of the retreating foe, who had retreated in the direction of Hancock and Sir John's Rim. We passed through Bath, taking the road to Hancock. Bath is the county seat of Morgan, containing some 500 inhabitants, noted for its springs and frequented during the summer by many visitors. It is a very handsome town, but greatly abused by the troops. It was the general impression of our troop that we would cross the river into Maryland. Col. Campbell of the 42nd Va. Inf. had taken the Sir. John Rim road coming upon the enemy at the railroad bridge. He had quiet a brisk skirmish routing the enemy and destroying the bridge. About the middle of the afternoon, we halted in front of Hancock. Col. Ashly was sent over under a flag of truce for what purpose I know not.

A brisk cannonying was kept up for several hours, which was responded to by the enemy. Several shells busted in streets. The yankees could be seen distinctly upon the opposite side.

Late in the evening we moved back some two miles and encamped. It commenced snowing again. The wagons did not reach us until after night. Distance ___ miles. (Whole 542)

---

2 David Hunter Strother was a Virginia native who joined the Union Army. He was well-known before the war as an artist and writer for *Harper's* under the pen-name "Porte Crayone."

Tuesday 6th

Remained in camp all day. Several men belonging to my company were from this county. I gave them permission to visit their friends. No news.

Wednesday, 7th

A day never to be forgotten by the Army of the Valley. Our wagons were packed and left camp early. The snow was some 6 inches deep and freezing cold. The roads had become so slippery as a mill pond. The whole army began to fall back. The wagons went in front and it was until late before they had passed. Genl. Jackson stood at a little hill and superintended the affair. It was dark when we passed through Bath. It was almost impossible for the horses to get along. The artillery was almost pulled by hand. I have seen every horse at a plce [sic] down at least once. Many of the horses were killed by falls upon the road. A quanity of the baggage was abandoned, but on and on we went until midnight when we reached Ungers Store, where we had encamped a few days before. It was almost impossible for the men themselves to travel. I have seen half of a company slide down at once upon the roads. Our field officers were compelled to abandon their horses.

When we reached camp, we had not one half of our regiment. The rest worn out and almost perished had stopped along the roadside and built fires and slept til morning. Many of the poor fellows never awoke again. While a number had their bones broken by falling upon the ice and a number were frost bitten.

When we reached camp, we found the Wagoners had built large fires and pitched our tents. They had procured some brandy of which all partook, being the first I had drank in four years. We were soon warm and enjoyed a smoking supper. Soon forgot our hardships in slumber. Distance 1_ miles. (Whole distance 560 miles)

Thursday 9th

Remained in camp today. Visited in company with Col. Harman, Michal G. Harman of Augusta. The weather has moderated and snow melted, making the camp very disagreeable, which was in low flat and in no order.

Friday 10th

Col. Harman received a leave of absence for 20 days, leaving me in command of the regiment. We changed camp some two miles further in direction of Martinsburg.

Saturday 11th

Furloughs were granted today to two men from each company. Orders were received in regard to the reorganization.

Sunday 12th

Received orders to be ready to move tomorrow at daylight. Nearly all the snow had melted and the ground was now frozen, which made it much better. Rations were cooked tonight for the morrow. No one had an idea where we were going.

Monday 13th

At dawn we moved off in direction of Romney. The men were very anxious to return to Winchester. We encamped late in the evening on the south bank of the North River. One and a half miles from Bloomery. Two days march was a very had one. Many of the men gave out on the tramp. Distance 18 miles. (While distance 578)

Tuesday, 14th

Moved early, the bridge being incomplete so we had to wait until it was finished. The weather was intensely cold and the men could hardly move when we marched off. We learned the enemy had evacuated Romney and destroyed much of the property. We encamped late within 8 miles of Romney, traveling some 14 miles. The country though which we passed had been totally destroyed by the enemy. Nearly every dwelling and barn had been burned. We were ordered to move at daylight next day. (Whole distance 592 miles)

Wednesday 15th

In compliance with the order issued last night, we moved at day light. It commenced raining and the roads were nothing but slush. At the junction of the North Western Va. Turnpike, a large tannery had been destroyed by the enemy and was still burning along the road. We reached Romney about 4 o'clock. We counted some 30 houses which had been burned along the road. I found quarters in houses for my men. The enemy had retreated toward New Creek. The town was in a delapidated condition. Nearly all the fencing had been burned and many of the buildings destroyed. Distance 8 miles. (Whole 600 miles)

Thursday 17th

Visited the yankee camps which were arranged in fine style. They had built winter quarters, expecting to remain all winter, but Jackson has frustrated their plans.

Friday 18th

The town was searched today and a large amount of store was found which the enemy could not destroy and they dedicated it for their own use. Dr. Snydern medicine was seized by Dr. Beveard of our regiment. The old gentleman became very indignent and in consideration of his daughters, I got them returned to him. This I gained the good grace of them.

Saturday 19th

It has rained every day we have been in town. The streets are very muddy, almost impassable.

Thursday 24th

We started this morning for Winchester. Genl. Lonig[3] was ordered to remain. The roads are very muddy. As we passed Lonig's troops, there seemed to be an ill feeling between the two commands. Genl. Lonig had been commanding a division and he did not like to be commanded by Genl. Jackson and each army sympathizes with their Genl. We marched some 15 miles and encamped. (Whole distance 615 mi.)

Friday 25th

We passed Mr. Blue's today. His house had been destroyed, his stock killed and an old gentleman who was in charge of the place was shot and afterwards burned in the dwelling. Hogs, sheep, cattle and poultry could be seen killed at every dwelling. We counted 34 homes burned on this road. This was the enemies try to eliminate us with fire and sword. God will not possess such fiends, such conduct, does their cause more harm than good. It only will make us more determined to seek revenge. We encamped some 16 miles from Winchester, making a distance of 15 miles. (Whole 630 miles)

Saturday 26th

We encamped today within four miles of Winchester and within 100 yards of my birth place. I visited John Jupton's this evening, where I got a handsome supper. (Whole 642 miles)

Sunday 27th

Visited home today. The wind last night blew our tents down. The weather is very cold.

Wednesday 30th, Camp Zollicoffer

We move to Camp Zollicoffer 4 miles north of Winchester on the Powatan Road, where we expected to put up winter quarters. Distance 9 miles. (Whole 651 miles)

---

3 Major General William W. Loring

February

Thursday 1st

We staked off our ground for winter quarters and commenced clearing the ground. The men are in fine spirits upon their return. On the march they censored Jackson and everybody else. A general dissatisfaction prevails in regard to the campaign. They could not see the object or the benefits of it. Our hospitals are filled with invalids from its efforts, our teams broken down, our horses thin and our baggage lost. I hope that something may show us that it is not labor lost. I am contented and willing that the future may develop the design.

Wednesday 7th

Men are busy at work with their quarters. The sound of the axe is deafening to the ear and the wood fastly dissappears under their strokes. The weather is agreeable. Col. Harman returned yesterday. The men are beginning to re-enlist and receiving the bounty and furloughs.

Monday 12th

Genl. Lornig has fallen back from Romney and destroying a large amount of his transportation and baggage, contrary to orders for which he has been placed under arrest. Genl. Jackson has tendered his resignation. The troops are very anxious it should not be received. A great deal of enmity has grown up between the two divisions. Fights are of frequent occurence in town when they meet. Each advocating the cause of their General.

Many of the huts are done and many under way, which will soon be completed.

Wednesday 14th

The rumor has reached us that the militia under Col. Shingendiver had been defeated at Bloomey Furnace (some 15 miles in advance of us) by the enemies cavalry.

Nearly all the huts are complete which are quite comfortable. They have covered them with clap boards. Many of them have plank floors. My company have hauled stoves from home. We are encamped in a fine neighborhood and favored with many pleasant visits.

Thursday 15th

The report of yesterday confirmed. The militia had been attacked by a body cavalry and most shamefully routed. Many taken prisoners. The others flying like sheep before them. Many never stopped until they got home. Among the prisoners was Col. Baldwin, Col. of the 31st Va. Militia. Some of their baggage was captured. They fell back to Powtown.

Tuesday 20th

The Yankees reoccupy Romney and it is rumored that Genl. Banks is preparing to cross the Potomac. The men are now enjoying their cabins and seem to be much at home, as they could wish. I received a letter from Farmington, the first I had received since I left Harpers Ferry.

Monday 26th

The men are reinlisting very fast. Some recruits are arriving and our army is increasing. About one third of our army is off on furlough. Lt. Maskell cashiered and all pay due him stopped.

Wednesday 28th

Held an election in my company for lieutenants. Lt. Fletcher was promoted to 1st and M. S. Brown to 2nd.

MARCH 1862

Thursday 1st

A long roll was beat, wagons packed and were ready to move. This was quiet unexpected to us and many thought we were going to march through our town. At 4 o'clock we moved to Glazes Woods within 1 mile of town and encamped. The troops enjoyed their labor but a short while. They had worked hard and took some pains in building their quarters. While they joked about them, many a love tale has been told by them as they sit before their hearths to enjoy the bright huge fire while blazed from the hearthstones. All this was now abandoned. (Whole distance 646 miles)

Wednesday 2nd

A dark cloud seems to hang over us at this time. The report of the evacuation of Manassas. The fall of Fort Donelson, the disaster at Roanoak Island and our expected evacuation of the lower valley seemed to fill everyone with fear and suppress the spirits of the soldiers, especially those of this army who expected in a short time would leave their homes to the mercy of a merciless foe, yet the men are reenlisting very fast and are devoted to the cause.

Our army is small undeciplined by inactivity and thinned by leaves of absence. While the enemy have recruited and decipined and equipped a second grand army of countless hosts, expecting to crush the rebellion in 60 days, which is the pledge of Mr. Seward, Sec. of State, to foreign powers. They are taking advantage of our weakness and pressing us upon all sides.

I have all confidence in a few months our army will be thoroughly organized, when we will turn around upon our foe with such vengeance that they will recoil before the gallentry of southern arms.

Thursday 3rd

Quite busy in Winchester hauling commisarys, quartermaster and ordinance stores, all seeing the heavy ordinance which is mounted on redoublets about the town and sending it to the rear. At the same time hands are employed in repairing the fortifications, which were constructed under Genl. Johnston last June. This is done to fill the public mind. I learn from good authority that the army is going through the same process as Manassas. It is with heavy hearts those of us from this section, witnessing the preparation, knowing in a few days the vandels will soon occupy our town and insult and restrain our families.

Friday 4th

About three this evening the drum rolled us to arms, our wagons were soon packed and sent through town on the Staunton Road. We were soon formed and marched out on the Martinburg Turnpike. We were placed in line of battle around three miles below Winchester. The fencing was ordered to be town down, artillery placed in position and skirmishers displayed. We expected any moment to see the enemy advancing. I never have seen men more eager for the conflict than this little army of a few hundred men. Nearly all of them were from this part of the valley. Here was their families and they would have died rather than to have been driven from the field. We remained in line until late. Then our wagons came up and the men bivowaced in line of battle. The enemies advance guard had come within two miles of us and then retired. Great excitement among the citizens. They, like us, expected any hour to hear the clack of arms in a deadly conflict. Many of the old gray headed fathers and boys came out to assist us with shot guns and etc.

Saturday 5th

We moved off this morning to a fortification on the top of Schultz hill, where a detail was made up to cut the timber off the hill and barracade the way. We were again marched back to our line of battle. (Whole distance 648)

6th, Sunday

We returned to our camp, still busy in removing stores from the town. The citizens are very uneasy. They, too, anticipate an evacuation. (Whole distance 650 miles)

Thursday 10th

Our wagons were loaded early this morning and pulled out in direction of Strasberg. We remained upon the ground until late, when again we were placed in line of battle where we were on Saturday. We remained here an hour or more when we fell back some 400 yards and then falling back to the fortifications. It was now nearly sunset. We could see the enemies skirmishers advancing some 2 miles off.

At dusk we marched off. No drums were allowed to be beat. At dusk we pased through our town. The citizens thronged the streets, many of them in tears. We expected to meet our trains near Winchester, but they had gone beyond Kernstown, where we halted and cooked our supper. My company was much depressed by leaving their families and friends. The citizens were leaving Winchester all night. A council of war had assembled and decided upon attacking them after night, but the train had gone to far to the rear. Distance 656 miles.

Friday 11th

We slept but little last night. The road was thronged with citizens all night fleeing from their homes. After breakfast, the wagons were repacked and started about sunrise. Our feelings were great as we turned our backs upon our homes, perhaps never to return. It is a lovely day. The trees have begun to put forth and the green herbs to carpet the earth. While the groves are made melodious with the woodland songsters.

This valley is the handsomest portion of our state. It looks as if the hades of war had never visited it, not a fence has been disturbed or a field overrun and in spite of the surrounding scenery, we were sad. Sad indeed to leave kindred loved ones and all that is dear unprotected in the hands of a foe. We knew not what would be the issue, everything dark around us, but we were soldiers bound to conquer, not only the foe, but the circumstances and mere necessity yield to fate. We passed through Newtown and Middletown and encamped on the south side of Cedar Creek. Marching a distance of fifteen miles.

General [Nathaniel P.] Banks forces marched into Winchester this morning at 3 o'clock in two colums, one by the way of Martinsburg, the other by the way of Beryville. First feeling their way in by a line of skirmishers, encirculing the whole town, then came the colum marching through every street with colors flying and bands playing, looking more like a set of boys dressed for an excursion than like soldiers. It surely is one of the best equipped armies on the continent with everything needed, perhaps we will fall heir to part of their equipage. Anyway, we were contented and rest upon the assurance that these sunshine soldiers will melt away before the front of the true southerners in his homespun suits.

Jimmy, my brother, only 12 years of age, came out after the enemy had possession of the place. They passed him through the lines as a country boy hunting his stock and when their cavalry were in close pursuit of several men who belonged to the continentals, he told the pursuers that a body of Stonewalls men were in Ambush, which caused a halt and enabled the pursuited to get ahead of them. Whole distance 671 miles.

Saturday 12th

Remained in camp all day. Nothing of importance occurring. A man by the name of Hodgson had come out from Winchester, who was supposed to be a spy. Genl. Jackson sent Lt. Smith and me to Woodstock to arrest him. We arrived at Woodstock at midnight, raining all the time.

Sunday 13th

We arrested Mr. Hodgson early this morning and started to return to the army, but on our way met the army falling back. They encamped at Narrow Passage, 4 miles from Woodstock, making a march of 18 miles. We were ordered to turn our prisoner over, which we did, and reported to our regiment. Still raining. Whole distance 689.

14th

Remained in camp. Men were busy washing and fixing up for continuation of the march. The enemy are slowly following us.

15th

Left camp this morning passing through Edenburg and encamping near Hamkinstown at Camp Buckhanan, not named after the officer who commanded the Merrimac. The distance marched 8 miles. Whole 697 miles.

18th

Commenced drilling and trying to reorganize our companies. We could receive but little news from Winchester. It is rumored that they are falling back and intend to evacuate Winchester. Genl. Johnston has fallen back to the Rappahanoc and it is thought that they intended an attack upon him.

Thursday 20th

We struck tents and moved our encampment to Rudes Hill, passing through Fort Jackson. Reached the encampment late. Rained all day. Very disagreeable. Marching distance 8 miles. Whole 705 miles.

Friday 21st

Appointed on court martial. Reported to Genl. Garnet for instructions. The Court was relieved and we were ordered to be ready to move. Late in the afternoon, we started in direction of Winchester and encamped near Hawkinstown, a distance of 6 miles. 711 miles.

Saturday 22nd

Left Camp early this morning and encamped at Ceeder Creek. A distance of 23 miles. The men nearly broken down by this march. We were informed all along the road that the enemy had evacuated Winchester. We were in high hopes of seeing our friends on the morrow. 734 miles.

Sunday 23rd

The sun as it arose this morning found us on the march. Everybody we met informed us that the enemy had left Winchester. We were in high hopes of seeing our friends and hearing from them the tales of the aggressions they had endured under Lincoln troops. Many were the speculations ventured upon. I fully expected to get home tonight and enjoy a good supper and visit my lady friends, nor was I alone in this vain expectation. Every man in my company fully expected it. We even made bets on it.

We passed through Middletown and Newtown. The ladies thronged the streets to receive us and sending with us their best wishes.

When we reached Bartonsville 5 miles from Winchester, we heard the enemies guns but the cavalry who were seeking the rear, told us that it was only the rear guard of the army, that the main force had left. We were halted in a trip of woods at the edge of Kernstown and ordered to load our pieces. I was ordered to deploy as skirmishers and to advance towards Neils Old Dam. I advanced a mile and a half as directed, when General Jackson and staff came up, he ordering a part of my men as guides, who were well acquainted with the country and I rejoined my regiment, which was then on the extreme right, which rested upon the Valley Pike. Cannonading increased. Several batteries was placed upon our right. About 6 in the evening the left of the line became engaged, musketry was very heavy. It was one continuous war. We were ordered into the engagement, which was on the extreme left some three miles from where we were on the right. The fire of musketry still kept up and grew closer and quickened our pace.

At sunset, we came in sight of the field. The 4th, 2nd, 33rd and 27th were falling back before the advancing enemy.

We tried to rally our troops, who had been engaged, but of no avail. We counter marched in the face of the enemy and took position behind a hill to cover the retreat of our whipped comrades. They soon advanced upon us. We opened them upon a brisk fire. We were repulsed, but rallied and returned and were repulsed again and again rallied and returned. By this time our annumition was well neigh exhausted. The 5th Ohio had attempted to flank us, but we repulsed them three times, they leaving their colors upon the field.

It was now getting dark. Hundreds of rifles flashed in our face. Still we stubbornly disputed the ground for this was the place we sported when young. This was our hunting ground and the scenes of our fishing excursions in schooldays and in the sight and sound of the town of our nativity, where now our friends were waiting to greet our return. Could we dissapoint them and permit the enemy to remain possessor of the field. With this determination, we rallied again being the third time and throwing ourselves upon the front of the advancing foe. The 48th Va. was brought up upon our left, discharging their pieces in the air. They hastily retired. Yet we opposed them until our ammunition was exhausted and the enemy had been reinforced. We were now compelled to retire, which was done in good order. We had so stubbornly resisted them that they failed to press their partial success.

I remained behind with what few men I could collect and tried to protect the wounded. We repulsed the cavalry several times and succeeded in getting off a number of the wounded.

The soldiers were scattered all over the county. Fire was built to guide them. We encamped just beyond Newtown within three miles of the battlefield. Genl. Jackson was on the side of the road cheering the men on calmy and seemed perfectly satisfied with the result.

As soon as I got to the fire, I began to collect my men together. I found a number missing for which no one could account. I was then ordered to Newtown to make arrangements for a speedy removal of the wounded for which we knew the enemy would press us tomorrow. I found a number in the hospitals and many that could not be removed. I returned to camp and reported to the General. I cannot speak too high of the courage and skill of our Col. W. H. Harman. He is brave, cool and collected on the field, always leading his troops and and seeking the most exposed portion of his command. Major Kolnir too behaved with great gallentry. My men fought well. It was for their own homes. Many of their parents and friends had collected where they could overlook the battlefield, offering to the good of battle their feeble and heartfelt prayers in our behalf, but they, like us, were doomed to sad disappointment. Our forces only numbered upon the field 2300 men, while they had actually engaged 10,000 men. It was not gallentry but brute force that gave them the hard earned field. Distance 700.

Monday 24th

At sun rise we were up and moving, passing through Middletown and halting on Ceeder Creek. Commenced cooking when a shell bursted in our encampment. The enemy were advancing in full force. In a moment everything was in the wagon and we moved off, owing to the skillful maneuvering of Genl. Garnett, we saved every wagon, and re-checked their advance as to continue our retreat in perfect safety. 6 were killed and 10 wounded in the 27th regiment today. We encamped at Maxwel Passage which we reached at 11 o'clock P.M. Many of the soldiers fell down perfectly exhausted upon the roadside, unable to go farther. In 3 days we had marched 70 miles and engaged the enemy and no one could stand it any longer.

But few of our regiment was able to reach camp tonight. 18 miles. Total 778 miles.

Tuesday 25th

Remained in camp. The stragglers who could not get in last night were coming in all day and some who were supposed had been left upon the field of the 23rd joined us. Our loss is estimated at 520 killed, wounded and missing, while the enemies is from 1500 to 2000. A flag of truce was sent to bury our dead, but were fired upon by the enemy. Our friends at home were not permitted to visit the battlefield, which caused a great deal of anxiety, but few families in Winchester but what had some relative in the engagement and was not permitted to know their fate.

Many censored Jackson for the engagement as being very rash and a want of skill, but I learned from him that he was promptly ordered to make the attack care not what were the odds, as the enemy had left Winchester to reinforce McLelland [George B. McClellan] who intended to make an attack upon Richmond, which if done would have sealed our fate for but a small army protected our capital. Several brigades were on the march to reinforce McLelland and had advanced as far as the Shenendoah, but were ordered back when Jackson made the attack. This foiling their intention. The loss of our regiments is estimated at 61 killed, wounded and missing.

Wednesday 26th

Moved to Camp Buckanan. The men are very much worn out by the hard marches they have lately made, yet they are in good spirits and perfectly contented with the result and boast of their courage. It was surely the hottest contested conflict and against greater odds than any battle of the war. DIstance 7 miles. Whole distance 785 miles.

Sunday 30th

We resumed drill and tried to reorganize them. Men beginning to be pretty well rested. The enemy have advanced as far as Woodstock. [My brother] Billy was taken very sick today and sent to hospital.

April 1862

Wednesday 3rd

General Garnett was placed under arrest by Genl. Jackson with the charge of cowardice and neglect of duty upon the engagement of the 23rd and Chas. H. Winder[1] was assigned to duty. Every man in the brigade was attached to Gen. Garnett and complimented his gallentry. I saw him on the field and noticed that he was among the last who returned. This was quiet unexpected and all regret his departure. Never was troops more attatched to a man than our brigade to him.

The enemy advanced as far as Edenberg. We were thrown in line of battle and at dark fell back to Hawkinston, a distance of 6 miles where we encamped. 791 miles.

Thursday 4th

Fell back to Rudes Hill, 5 miles from where we encamped last night. No news of importance. 796 miles.

---

1 "Charles H, Winder." Refers to Charles Sidney Winder who succeeded R.B. Garnett in command of the Stonewall Brigade. Born Maryland, a nephew of Confederate Admiral Franklin Buchanan. West Point, 1850. Received an early promotion to captain for heroism during a hurricane at sea. Resigned April 1, 1861 and was appointed Confederate brigadier general March 1, 1862. Praised by Jackson who was chary with compliments. A later entry reports Jackson concerned about Winder's illness. Though ordered not to, Winder went into action while seriously ill and was killed by a shell on August 9, 1862

5th

We were ordered to report the number of men required to make our present number reach the maximum, which was 125, in order that the militia could be drafted. My company was unfortunately situated. The enemy having possession of our county. We had not the resource that others had similarly situated had to receive recruits. I require 27 to make my present number reach the maximum. The militia does not like the idea of being drafted. They would prefer remaining in their present organization.

6th

Militia were drafted today. Those for my company were taken from the Augusta and Shenendoah County.

Friday 7th

Two of our companies reorganized today. Co. E. and C.

Commenced snowing. Men have no wood. Compelled to burn rails.

Saturday 8th

My company reorganized today. I was reelected Captain. Lt. Fletcher, 1st Lt., Lt. M. S. Brown 2nd, and J. W. Jones for 2nd Lt. Each was unanimous. Co. K also reorganized. Captain Smith, Geo. W. Kuntz, J. R. Mesmore and M. Formey were the officers unanimously chosen.

Sunday 9th

I attended church in New Market and visited Uncle John's. Found his family a very pleasant one. All quiet. No news from the enemy.

Monday 10th

Ordered on picket at Hawkin's Town, 5 miles down the valley. Part of my regiment was sent on outpost this evening. Genl. Ashley is in command of the pickets. (Whole 801 miles)

Tuesday 11th

Relieved and returned to the reserve. All quiet along the line.

Wednesday 12th

The long roll aroused us this morning early. Our wagons were immediately loaded. The enemy commenced cannonrading. The made a dash upon the bridge, killing Ashley's gray horse. We were beyond New Market, where we cooked our ration and started again on the march. Encamping at the Big Springs. A distance of 16 miles. Total 816 miles.

Thursday 14th

Took up the line of march early. The enemy still pressed us. Passing through Harrisonburg, we took the Swifton Gap Road and encamped 6 miles from the town. It had rained all day and the roads were very disagreeable. Many of the men thought we were evacuating the valley and deserted. Several militia from my company went home. It continued to rain. Total 833 miles.

Friday 15th

Continued to rain. Passed through McAysville, crossed the Shenendoah River. Very disagreeable marching and the men have no tents but few officers have even got a tent. Distance 16 miles. Total 849.

Tuesday 19th

Sent on picket today. Still raining. Men suffering very much from the weather, exposed and without tents.

Friday 22nd

The regiment was reorganized today. William S. H. Baylor elected Col., myself Lt. Col., W. J. Williams Major.

Saturday 23rd

Our camps had become so disagreeable that we were compelled to move upon the hill. The 16th Va. Infantry was sent over from the army of the Potomac and joined us today.

Monday 25th

An election was held to fill the vacancys caused by promotion. Lt. Fletcher promoted to Captain, W. S. Brown to 1st. Lt., J. W. Jones to 2nd. Lt. and J. W. O. Funk, 2LT.

Tuesday 26th

We moved our encampment to the foot of the mountain, a distance of 4 miles. Here the news reached us that Genl. [Richard S.] Ewell was marching to our reinforcement. This cheers the men. (853 miles)

Friday 29th

Part of Genl. Ewells forces reached us today. This is cheering the brave troops of Kennston.

Saturday 30th

Returned to our camp quiet unwell, unable to remain in camp. Ordered to be ready to move at sunrise tomorrow. (Total 857 miles)

May, 1862

Wednesday 30th April

Left Camp this morning at 4. Crossed the river and lay quiet beyond the bridge until late in the evening when we took the Port Republic Road. A chilling rain fell all day, making the roads almost impassable for men, much less the wagons. The wagons were stalled and broken down for miles along the road. We encamped in a woods some 14 miles from where we started. Not one half of the men reached camp. Our wagons could not come up. The men were compelled to remain without anything to eat. No clothing. And it rained all night. I never suffered so much in my life. 371 miles.

Tuesday 1st

Our wagons came up this morning. After cooking, we repacked the wagons and took up the line of march. Continued to rain all day. The roads have no bottom. One half of the regiment was detailed with the wagons and cut new roads. Nearly all the way was compelled to throw many of their things away. Marched 12 miles and encamped. I never spent such a disagreeable day. 383 miles.

Wednesday 2nd

Rained all night. Our commisary train could not reach us. Had to corduroy the road in many places today. A number of the horses died. The men threw away all their baggage except what they could carry, and it was very difficult for the wagons then to get along.

Reached Port Republic late in the evening, a distance of ten miles. No sooner than we reached the place, the yankees made their appearance on the opposite side of the river. No one has an idea where we are destined. 893 miles.

Thursday 3rd

Took up the line of march this morning, crossed Brown's Gap in the Blue Ridge and encamped at the foot of the mountain in Albemarle [County], a distance of 17 miles. It had cleared off and the weather was magnificent weather and a romantic scenery from the top of the mountain to look over the outstretched valley of Virginia for the liveliest spot in the state and clothes in the robe of spring.

The Valley of Italy can not be more enchanting to the eye and the circumstances make it more charming. We're leaving it perhaps for the last time to be pillaged and trodden down by the foe.

Many thought we were going to Richmond. In fact, all was of the same impression. Everything was dark around us. Our Army had fallen back to Richmond and the enemy were entrenched within sound of her bells and in sight of her domes. The fate of Richmond seemed almost inevitable and if she fell then, we would be compelled to return beyond the limits of our state. Indeed it was a dark hour. Even the most hopeful feared and trembled. How anxious was our last look upon the valley. Think as we did as we reluctantly turned our back upon the first scene for the last time. Many of us many fall upon the field of battle if successful.

The valley of Albermarle, now spread out before us as far as the eye could reach, not much less lovely than the valley we had just left, but much less interesting than the one of our nativity.

Everything had advanced more than it had on the west side. Not a fence has been distributed, not an army has yet marched through this part of the country. 910 miles.

Friday 4th

Marched to Mechum River Depot[1] where we encamped a distance of 13 miles, passing through a pleasant country. Our sick and unable took the train for Staunton. We were gratified when we knew we were going back to the Valley. 925 miles.

---

1 Important rail station and bridge between Charlottesville and Staunton.

Saturday 5th

Marched to Afton Station 10 miles. Several regiments had taken the train and gone to Staunton. Our regiment was in high hopes of soon seeing their friends. 935 miles.

Sunday 6th

Took the train this morning at 4, passing through Waynesboro and encamped within two miles of Staunton. The ladies sent us several loads of provisions, enough to supply the regiment.

Col. Baylor has got permission to take his men near town to encamp. We passed through town which was thronged with the friends and relatives of the men and encamped on the west edge of town. Load after load of provisions was brought in to camp. The regiment mostly from this county were permitted to visit their friends for 24 hours, all day our camp was thronged with the wives and sweethearts and mothers of the regiment. We spent a delightful time of it. Distance 941 miles.

Monday 7th

Left camp early this morning taking the Monterey Turnpike, We had anticipated spending a few pleasant days in Augusta [County]. We passed through West View. Co. F. had been formed from this town. They were permitted to remain a few hours to enjoy a dinner prepared for them by their friends.

Encamped in Buffalo Gap, a distance of 16 miles. Here we came up with Genl. Johnston's command, which had fallen back and the enemy had advanced this far, but upon approach, had fallen back with some confusion, leaving their baggage and camp equipment and some few prisoners. 955 miles.

Tuesday 8th

Marched across the Shenandoah Mountain this morning. A fine view to look over the Valley of Cowpasture. Our march was rapid in order to overtake the enemy. On the heights one mile this side of McDowell, the enemy had taken strong position. Genl. Ed Johnston.[2] was in command of the advance. About 4 P.M. the engagement opened. We were some 4 miles from the field. Throwing off our knapsacks, we hurried forward to the relief of our friends. The contest grew warmer and warmer.

A number of wounded men were being brought of the field. When within a mile of the engagement we were ordered back, returning to our wagons. Tired and worn with the duties of the day, we were soon asleep, only to be aroused by the long roll. We were ordered upon the field. We pushed on with all the speed that worn out men could.

The field was on a high hill. We ascended it, meeting a number of wounded being brought off the field. We were then ordered to lay down. Col. Gibbons of the 19th VA. Regiment was killed. Genl. Ed. Johnston wounded. About a mile from the field, we could see a host of the enemies campfire. We had repulsed them after a desperate effort.

The 12th Georgia regiment seemed to suffer more than any other regiment, losing 5 captains out of 10. Distance marched 33 miles. 9___ miles.

Wednesday 9th

We laid in the battle field until 2 o'clock P.M. when we were ordered to return to our wagons. Up early and on the march, the enemy had fallen back, abandoning their stores and equipage. Here was the first time I met with the boys from Marion. All were well and in fine spirits. The enemy left a number of their dead and wounded behind. Several pieces of artillery was captured. General Milroy was

---

2 Major General Edward Johnson

in command. It was impossible for us to get our artillery into position yesterday. We had been for some time retreating from the foe, and this was the first time we had followed them and it was enjoyed.

Our loss on yesterday was 42 killed and 250 wounded. The 12th Georgia Co. has won for its gallentry and imperishable fame.

We passed through McDowel where several houses had been burned by the enemy. We encamped some 5 miles beyond McDowel, a distance of 14 miles. Total 1002 miles.

Thursday 10th

We followed up the retreat of their routed army with as much speed as our worn out troops would admit. Along the road their retreat was marked by the graves of their dead. We took the Franklin Road, 3 miles from Monterey.[3] We were informed by Captain Scheetz that the enemy had ambushed one regiment some two miles ahead. We double quickened it for some 2 miles, when the news came that they had retired. We encamped on the head waters of South Branch. 1016 miles.

Friday 11th

When we got within 2 miles of Franklin, we were halted and two companies from my regiment sent out as skirmishers. After the skirmishers had advanced some half a mile, we followed as soon as we made the turn into a flat some two miles in length.

The enemy opened upon us with a battery of two guns. We were immediately thrown into line, when the enemies line moved off and took refuge around Franklin. Our regiment was ordered upon the top of a high and bushy hill to deploy as skirmishers.

It was with great difficulty we could get along. We drove in the enemies pickets for nearly a mile. They had set fire to nearly everything and we could not see over a hundred yards for the smoke. When within a mile of Franklin our pickets had a brisk skirmish.

---

3 Monterey

Driving the enemy before us, we remained here a short while, when ordered nearly a mile to the right, where we lay so close to the enemy that we could hear them talk. Yet the smoke was so dense we could not see each other. I attempted to pick off their cannoners with sharpshooters, but with no effect.

After dark we went two miles to the rear where we bevowaced for the night. We had several wounded among the number was Captain James Gibson of Company H. 1028 miles.

Saturday 12

Went on a picket this morning, remaining undisturbed until 3 when we were ordered back to camp.

The enemy shelled the woods furiously around us, but without effect. The General published an order this morning announcing a victory and asking us to devote the day to prayer and thanksgiving to God for the blessing visited upon our army. Late in the evening we fell back some 5 miles and encamped. 1035 miles.

Sunday 13th

Left early this morning encamping near McDowell. Rained all day. Disagreeable marching. Many of the men sank exhausted on the road side and a number of wagons were abandoned on the march. The soldiers have been amusing themselves reading the letters taken from the enemy. I have never until now thought thare was as much vulgarity.

The citizens had prepared a dinner for the 10th regiment from this place, yet there was a gloom. They had lost the lamented and much beloved Gibins, who had fell on the 8th. He was a noble officer and generous to his men. This has completed a circle of over 398 miles in less than 30 days. And fighting one battle and chasing the enemy over 40 miles. 1115 miles total.

Wednesday 21st

A beautiful day. What grain has been undisturbed is approaching harvest. The trees are laden with fruit. We met with a very enthuestic reception by the ladies of New Market. We took the Luray Turn Pike and encamped at the foot of the Masullens Mountain, 4 miles from New Market. Before we reached New Market, we were halted over to permit the head of the colum to countermarch that had passed beyond the junction of the road, which we were to take and it was rumored the protestation of General Jackson, sent to General Johnston, had no effect and that General Jackson had received prompt orders to proceed at once to Gordonsville by Sperryville. This was generally credited by the army. The men were sadly dissapointed when they found that the army had taken the Gordonsville Turnpike.

I visited New Market this evening and found the citizens highly delighted by our army bring in the midst. The citizens had been very loyal and behaved with great credit during the short reighn of Uncle Abe. The ladies would not notice the Federal Officers or walk beneath the flag. The soldiers ever feel proud of their conduct of the women.

Thursday 22nd

Crossed the mountain passing through Luray and taking the Fort Royal Road. You can't imagine the extacy of the troops, when they know they were destined to the lower valley.

I arrived at Mrs. Jordon's, the mother of the A. S. General of General Beaurgard. We encamped 4 miles from the town.

While laying in my tent this evening in a daze, I imagined that I heard a bullet whislte and it struck me in the breast, passing through. This wore heavy on my mind, thinking it had foretold my fate, but being worried and worn out, sleep soon relieved my mind. 1139 miles.

Friday 23rd

Men were ordered to leave their knapsacks and blankets with the wagons. We marched in direction of Front Royal where the enemy were posted in small force. We found the roads blockaded, which led into the town. About an hour before sunset, the enemy opened with artillery upon our advance. A brisk skirmish of infantry ensued which lasted about twenty minutes. When the enemy were entirely routed and nearly all of them taken prisoners, with their artillery and stores. Nearly all the 1st Mo. regiment (Yankees) was taken prisoners by the 1st Va. Regiment (Confederates). From a hill just before we entered the town, we could overlook the valley as far as the Potomac. It was a lovely sight. The sun was just rising which added splendor to the already magnificent scene. We loved see our home, which was in a days march and we knew we would drive the enemy from it, yet it was a sad thought. No one knew who would survive or fall in the struggle and the thought of my dream still worried me, yet I know of no spot on earth I would rather sacrafice my life in defending.

It was dark when we drove the army from town. We passed through town and encamped just beyond on the Winchester Turnpike. General Shields with 7000 troops had occupied this country and he a few days before had crossed the mountain to reinforce General McDowell at Fredricksburg, which was the right wing of army that threatened Richmond and General Jackson had taken advantage of this movement and was throwing his whole force on the enemies and they, having whipped and driven back Banks right wing at McDowel and suddenly falling upon his left and routed it. The main body of Banks army is now at Strasburg. (1157 miles)

Saturday 24th

Our wagons came up this morning and the men have been busy for several hours preparing their rations.

Captain Sheetz was killed this morning and the news of his death spread a pall over the army. He had been universally admired for his gallentry by the whole army. He had done good service in Hampshire [County] while the enemy were there. We boasted not of our victory for we had lost our Sheetz. This was the sentiment of the army.

A large number of prisoners were brought in this morning by the cavalry who were still pursuing them. We had captured an enormous amount of clothing and stores.

We crossed the Shenendoah River and proceeded in direction of Winchester. This bridge had been burned by us in the retreat up the valley, but rebuilt by the enemy. They attempted to burn it in their retreat, but our cavalry had dashed upon them and prevented it. The road was lined with wagons artillery and camp equipment for miles and prisoners were brought in by large numbers. Our division left the road and turned toward Middletown. While General Ewell's forces went on in direction of Winchester, hundreds of prisoners and horses were brought by us. The enemy had fallen back from Strasberg and their train guarded by cavalry had been totally routed and the wagon train captured.

When we reached Middletown, we took the Winchester Turnpike. Hundreds of wagons loaded town with camp equipment, sutler stores and officers baggage lined the road.

While arms of all description were spread in every direction, a number of dead were laying around. The men amply supplied themselves with everything they could wish from the wagons. The sutler wagons were filled with everything that was nice to the taste. This was quiet a treat to an army from which such luxuries were excluded.

At Newtown the enemy had taken position to retard our progress. They soon retried firing their trains. We pressed on passing through the town. It was now night, but the burning train made it almost as light as the mid-day for several miles there was a continued line of wagons all in flames, consisting of pontoons, Qr. Masters Commisarry and Ordinance stores. At Bartonsville, the enemy were ambushed, firing upon our advance, which stampeded several regiments, which came neigh routing the entire army. Col. Grigsby was ordered to the front. I taking position on the right. After a brisk engagement for a few moments, the enemy were driven from the field, leaving their dead and wounded. I was there ordered to the front to follow them up as skirmishers. I sent forward Captain Fletcher's Winchester Rifles and Captain Kertz companies, both from Winchester and was well acquainted with the country. The enemy kept up a continual fire upon our skirmishers, but with little effect.

At Kernstown, they made a more bold stand than any place along the road, but soon retired leaving their dead and wounded on the field. I had their wounded carried to a shelter, where they were dressed by our surgeons. We arrived at Kernstown near midnight, a distance of 19 miles. Total miles 1176.

Sunday 25th

The men had become very much fatigued, having no rest or sleep. The skirmishers, many of them had become exhausted, wading through the tall grass on each side of the road. I asked the General to permit me to relive them with men more fresh, but he didn't, by replying "They do their work nobley".

We continued to press them and they falling back. General Jackson and General Ashley were up with the skirmishers all the while. We soon came in sight of their camp fire. I gave the General all the knowledge of the country I could.

The dawn of day found us close to Winchester. The General ordered me to take the hill which overlooked Winchester. By daylight a courier informed General Jackson that General Ewell was within a mile of Winchester on the Front Royal Road. After resting, the men for a little while, we again began to advance. The cavalry made a dash down the road, but a few shots from the skirmishers repulsed them. I was then ordered to take Boners Hill. The 27th and 2nd was ordered to my support.

No sooner had I got on top than the enemy opened a tremendous fire of artillery. I awaited with anxiety for orders to move forward, wishing to get in my native town. The order to charge soon passed along the lines and with a tremendous shout, we hurled on the foe, completely routing them. I was the first Confederate to get into town. We could scarcely pass along the streets for the ladies who came to welcome us back. We took a number of prisoners in town. Col. Baylor was wounded and his horse killed. I fired upon the yankee who wounded him in front of Wm. Oliver Browns and killed him. This was the first man I ever knew of killing. We continued the present. Frequently the enemy would try to check us but a few well aimed shots from our artillery would disperse them.

I have never seen our army as much demoralized as this. After following them some 9 miles from Winchester, we were ordered into camp.

I got permission and returned home. This is a day never to be forgotten. The enemy has been pressing us upon all sides for the past four months and our sky has been dark, but this victory will be hailed with the wildest enthusiasm throughout the Confederacy,

Upon our entrance into town, the ladies would rush to meet us between the two fires and we could not prevail upon them to return to the house out of danger. They had heard of our success at Front Royal and made a hasty preparation for our reception. Nearly every house in the town was thrown open for the soldiers.

The enemy had fired the buildings containing their medical stores but the ladies carried water and extinguished it. This saving a valuable amount of medicine. W captured a large amount of Sutler stores in the town and over 3,000 prisoners and enormous amount of clothes.

Banks retreated in direction of Martinsburg. Part of our forces went to Romeny. They crossed the Potomac at 3 o'clock today, making 27 miles in 6 hours. I found that the enemy had done father but little damage. My brother was confined in jail for some two weeks. The Marion Rifles and Morgans Continentals were permitted to remain in town. 1197 miles.

Monday 26th

Returned to the regiment today. Men were busy washing and cleaning up and talking of the victory of yesterday. General Jackson was no crazy or mad man now, as they would call him when worn out on a march. Many vowed that they would say nothing against him hereafter.

Wednesday 28th

Ordered to Charlestown, when within a few miles of the town, we were informed that the enemy had come up from the ferry this morning. I was ordered to send out skirmishers, which soon came upon the enemy. When within a mile of the town, they commenced shelling us. Carpenters Battery was soon in position and returned the fire. The enemy the returned to the town setting several buildings on fire.

We were then greeted by the ladies when we entered the town. We continued in pursuit until within a few miles of Boliver Heights. Then we returned to within one mile of town and encamped, marching a distance of 26 miles. Several ladies in town had been struck with sabers by the Federal Cavalry as they retreated. This is one of the most loyal towns in the state. Total 1213.

Thursday 29th

Visited town this morning. In the evening we were ordered to the front. General Jackson and Ewell had arrived with their forces from Winchester. We were thrown out as skirmishers on the right of Boliver Hill. The enemy were cannonading us, but their balls fell short. We had advanced nearly a mile and were then ordered back to our camp. 1223 miles.

Friday 30th

Returned to where were posted yesterday with the belief that we would have to storm the heights, but when we had advanced near the foot of the hill, we were ordered to fall back to our old camp. 1233 miles.

Saturday 31st

Commenced falling back. Took the road leading to Smithfield. When within a few miles of the town, the scouts informed us that the enemy were trying to cut our train into at Barkers Hill. I was hurried up to support the trains. When we reached Smithfield, we took the road by Brucetown. Commenced raining very disagreeable to the troops. We passed through Winchester and encamped near Bartonsville, marching a distance of 31 miles. It was very painful to leave my town the second time, yet it was the fate of war and we must submit. Many of the Sutler stores could not be carried off, so the General ordered them open for the troops.

Continued to rain very hard. We were compelled to leave some of our men behind who were unable to travel. I could not even get home to see my parents. Total 1263 miles.

June 1862

Sunday 1st

General [John C.] Fremont had been ordered from Franklin and General Shields from Fredricksburg to the Valley to destroy Mr. Jackson who had jumped on Banks and run him out of Virginia. Fremont had made his appearance near Strausberg and Shields was near at hand. We were compelled to make a forced match to prevent being cut off. Upon our removal at Strausberg, General Jackson remarked to General Winder that he felt very much concerned about him.

The report of artillery told us that General Ewell had already engaged Fremont calmly. We had hardly arrived at Strausberg when Shields made his appearance too. We lay in line of battle until dark when we went in camp some 6 miles from Strausberg. Continued to rain. Many of soldiers were compelled to remain behind unable to keep up. Total 1282 miles.

Monday 2nd

The enemy pressed us early. We were compelled to move early. The cavalry was routed and came dashing over us and tramping the men down. Our brigade was ordered into line. We soon checked them until our wagon train was safe, then slowly retired. It was very hot and sultry day and we had to march through wheat as high as our heads. Many of the men gave out. I gave a man my horse who was exhausted. Our men were so completely exhausted that we were compelled to lay in the woods for several hours before we could resume the retreat. We passed through Edenburg and camped near Hawkinstown, a distance of 19 miles. Total 1301.

Tuesday 3rd

Passed through Mt. Jackson and encamped near New Market. Shields[1] was advancing up the Devay Valley, while Fremont was at our heels. Rained all day. Total miles 1309.

Wednesday 4th

Remained in camp until late this evening, when we marched to the big spring and encamped. News reached today of our victory at 7 Pines and of the victory of General Joe Johnston. Our Star seems to shine brighter now than a few months ago and that our valor is equal to their number. 1318 miles.

Thursday 5th

Marched through Harrisonburg, where the men got the baggage they left when they went down the valley. The bridge being incomplete at Mt. Crawford, we took the Port Republic Road. Encamping on the bank of the Shenendoah near the town. I have never in my life seen the men as neigh exhausted, out of 450 men I arrived at camp with 15 men. The rest fell exhausted on the roadside. It was very muddy and almost impossible for men to travel. The troops have been marching every day for the past 30 days, but 4 days rest, and fought three battles and a number of skirmishers and marched over six hundred and twenty five miles.

This is more than men can endure. No wonder the poor soldiers fall down exhausted on the roadside, being much harder upon us for our brigade has been the rear guard. It was near midnight when we reached camp. 1342 miles.

---

1 Shields. Union General defeated by Stonewall Jackson at Cross Keys and Port Republic, June 8-9, 1862.

Saturday 7th

The commanders petitioned through General Winder in behalf of the brigade asking for rest for the men. Col. Baylor went home yesterday on sick leave. The men are busy in washing their clothes and making a general clean up. General Winder and myself intend visiting Boyers Cave.

Sunday 8th

I was in my tent this morning preparing to go to the cave when I heard the report of a cannon. I gave it no attention, thinking it was one of our guns. In a few moments, it was followed by another. Still I paid it no attention. It was no time until another was heard and one of my men who was washing in the river, busted into my tent and told me the yankees had crossed the river and had possession of the bridge. I stepped to the regiment and found that our wagon was unloaded and the men busy cooking. I at once ordered them to pack their wagons and fall in. In a few moments we was in line and moved down to the bridge. On my way, I met General Jackson who had just crossed the bridge and he ordered me to support Plogues Battery that was just moving into position. The enemy had just planted a piece on the opposite side of the river to sweep the bridge. General Jackson thinking it was one of his pieces, rode down to the bridge and ordered it to come up on the hill, but discovering his mistake rode off. In a few moments, a squadron of cavalry came dashing up to the bridge, when the General told them to go away, which they did.

Our battery was soon in position, which opened upon them. They fled, leaving their pieces: Several colums of infantry were advancing up a bottom on the opposite side and our batteries gave them their attention, firing some well aimed shots into their midst, forcing them to fall back. They tried it the third time and with the same success.

By this time, General Ewell had engaged Fremont at Cross Keys,[2] 5 miles in our rear. A tremendous fire was kept up all day. We expected to be ordered to his support, but we had to watch Shields.

---

2 Cross Keys.

At night the engagement ceased. General Ewell had driven him from the field and held every position of it. A number of prisoners were brought to the rear. Shields advance had encamped some 4 miles below Port. At dark, we crossed the river, passed through town and encamped a mile beyond, where a days provisions was cooked and ammunition distributed. Our loss today had been estimated at 500 killed and wounded. The enemies much heavier. Among the wounded was General Elsig [sic, Elzey?] and General G. H. Stewart. 1346 total miles.

Monday 9th

At daylight this morning, we crossed the river on a temporary bridge and advanced upon Shields, who was still on the Swifton Gap Road. We had advanced but a short distance when a shell was hurled into our midst from the enemy. I was ordered to support a battery on the left of the road and had hardly got in position when ordered to the left near the bank of the river, where I threw out skirmishers.

I was then ordered to charge a battery on a hill some 400 yards in our advance. When within a few hundred yards of the pieces, I was ordered to halt and await the arrival of the 7th ordered to my support. Col. Hays was in command of the left wing. We were then ordered forward under a murderous fire of casester shell and musketry. After wading through a pond nearly waist deep, we were then ordered to open upon the enemy, I found that I had to contend against a brigade which consisted of the 5th Ohio, 7th Indiana and a Pennsylvania Regiment which only four hundred men. I was determined to hold my ground as long as it was possible and sent to General Winder for reinforcement, that my men had exhausted nearly every round of ammunition and could hold their position but little longer.

A destructive fire was continued for nearly 60 minutes. My men were falling all around me. Having exhausted their ammunition, they stood still in line with fixed bayonets, waiting the order to charge. In the meantime, the 27th Va. Infantry supporting the center gave way. I was ordered to fall back twice before I did and then the enemy had nearly cut us off. I fell back to a barn some three hundred yards in our rear and reformed filling our cartridge boxes from those of our comrades who had been wounded or killed. By this time, their right wing had been turned by our forces and they commenced a hasty retreat.

We continued to follow them for some 3 miles when we were ordered back. Our forces which were on the opposite side of the river had crossed and burned the bridge and Fremonts forces soon had possession of the opposite bank, which overlooked the battle field, but he had come too late. Shields had them routed. Neither could he cross to his assistance nor I do not think he desired to cross much for he had a test of our courage the day before at Cross Keys where he left a number of the celebrated buck trails [sic, Pennsylvania Bucktails] on the field.

Fremonts artillery now commanded the road leading to Crows Gap the road our wagons took this morning, so we took a back road coming out on the main road some three miles up the mountain. I visited the battle field near dark to see if any of my men remained upon the field, but Fremont shelled us off. We reached camp about midnight, which was on the top of the mountain. The road to camp was so blocked up with artillery, cavalry and infantry, that it was difficult to get along. Lt. Anal of Co. I fell which leading his company. He was a brave and intelligent officer and none knew him but what admired him. His loss was severe to the regiment. The loss of the regiment was 115 killed and wounded. I took into the engagement 385 men. I never saw men fight more gallant then upon this occasion. 1362 miles.

Tuesday 10th

Busy all day removing the wounded and burying the dead. Several of the wounded men who were left upon the field were bayoneted and struck with muskets by the enemy when they took possession of the field.

Rained all day. Very disagreeable, but few of the men have tents. 7 prisoners were captured yesterday. Our regiment got the colors of the 5th Ohio.

Wednesday 11th

Marched down the mountain. Passed through Port Republic and encamped near Weyers Cave, a distance of 13 miles. Fremont having lost his path found Jackson and returned to Woodstock contented with his campaign, while Shields army had received such a shock that he was ordered to Alexandria to recruit.

They committed frequent depredations upon the citizens during their short sojourn. Blinkers duck were celebrated for theft not so much for fighting. 1375 miles.

Thursday 12th

Visited Weyers Cave, one of the natural curosities of Virginia. It is magnificent affair, frequented by many travelers and classed among one of the first curiosities of the continent.

Saturday 14th

Men have been busy cleaning up and seem to enjoy the short rest. General Whiting's division arrived today from Richmond. Major Dabeney[3] our A. A. General, preached for us today. General Jackson and his staff were present. Quite busy in trying to reorganize our regiments and in punishing delinquents.

Sunday 15th

Major Dabney preached for us again. General Ewell was also present. Then men are in excellent spirits. Several new regiments arrived.

Tuesday 17th

We moved this morning early in direction of Waynesboro, contrary to our expectations, for we had thought that we would soon make a descent upon the enemy some 8 thousand troops had been sent to reinforce us. We passed through Waynesboro and encamped at Rockfish Gap. While the most of General Whiteing's command took the train. 1390 miles.

---

3 Robert Lewis Dabney, Presbyterian theologian.

Wednesday 19th

Crossed the mountain into the lovely valley of Albermarle and encamped 4 miles beyond Mechum River Depot. The day being very sultry and roads dusty, it was very disagreeable marching. Distance 18 miles. Total 1408.

Friday 20th

Passed through Charlotteville. Great demonstration among the citizens on the appearance of the Stonewall Brigade. Facts were profusely distributed by the ladies.

We encamped four miles from the town near Monticello where lies the ashes of Tom Jefferson, who so prophetically spoke of the present. Men very much worried by the march. 1425 miles.

Saturday 21st

Encamped near Gordonsville. Men very much exhausted. Day dry and sultry. 1440 miles.

Sunday 22nd

Remained in camp today. Col. Baylor caught up with us. Had preaching. Men were quite busy in washing their clothes.

Monday 23rd

Marched three miles below Louisa Court House and encamped, making a distance of 15 miles. 1455 miles.

Tuesday 24th

Took the train and went to Beaver Dam. Remained here until late in the evening, when we marched in direction of Richmond. The roads were dreadful. The poor soldiers pulled one foot after the other with great difficulty, now and then plunging into a hole in the road waist deep. Many remained on the road side. The brigade was strung along for over 5 miles, finding it impossible to go farther.

I, with a few men who could keep up, built fires in the woods and remained the rest of the night. I have never seen men as badly used up as they are now. 1465 miles.

Wednesday 25th

Moved early this morning, thinking that we could get to the wagons to prepare something to eat, but the wagons were in our rear, so the men were compelled to toil on through the mud all day without having anything to eat since the morning before.

We encamped in some thick pines where hand machines were issued and we were ordered to leave the wagons and take three days rations. Not a soldier in the regiment but what knew the meaning of it. It meant to fight. The men were worried, but ready to do what was required and went to work cheerfully. 1480 miles.

Thursday 26

Moved early, about 10 P.M. General Stewart and his cavalry came up. About noon we could hear the booming of artillery and was informed that General Hill had driven the enemy from Machaneksville.[4] We were now approaching the enemies lines. Tomorrow we expect to be hurled upon his right wing. 1492 miles.

Friday 27th

A day of victory in the epoch of our history. We moved early, about noon came upon the encampment which the enemy had abandoned yesterday in the vicinity of Machenecksville. They were arranged with some taste. Here we lay for several hours now and we could hear the booming of the cannon, making way for the bloody engagement which was soon to follow. About three we moved in the direction of Gaines Mill, passing over a portion of yesterdays battlefield, which was covered with their dead and baggage. The

---

4 Machaneksville. Refers to Mechanicsville.

firing became very heavy and the rattling of small arms could be distinctly heard and we pushed forward with all the speed we could. We were soon in sight of the field.

The smoke could be seen, column after column rising towards the heavens for over _ miles in length. We pressed forward leaving our knapsacks behind. We soon crossed a swamp and came in full view of the field. General Hill and Lee were together. The enemy occupied strong position upon a hill called James Hill. The worn out troops seemed to welcome us. General Winder was in command of our division and Colonel Allen in command of the brigade. General Hill told General Winder that he could not get his men to charge, but that they would stand and fire as long as anybody and that he wished him to send a brigade upon the left that would charge the enemy.

The Stonewall was chosen for the alloted task. We were ordered to support a battery just placed in position, but soon moved to the left where we advanced through a woods in a swamp coming out into an open field just in front of the enemy. We were then ordered to charge a battery which was firing death into our ranks. It was now near sunset. We pushed on running over their lines of battle and attempted to get them to advance with us, but of no avail. We dashed on with all the speed we could, but at every step leaving many of our brave comrades on the field. We soon drove the enemy from their pieces, which were left in our hands, but not until our ranks had been dreadfully thinned by its well aimed fire. It was now dark but we still pushed forward until we came upon a swamp and was compelled to halt not knowing the roads and too dark to see them. We then fell back to the top of the hill and lay in line of battle all night. This was a glorious victory, but in achieving it, we had lost many brave men. Col. Allen, Capt. Shiver, Lt. Sheets of the 2nd fell. Our loss was 64 killed and wounded. Among the killed was Captain W. H. Randolph, a noble soldier and a gallant officer. Among the wounded was Captain Fletcher and Lt. McKamey and Lt. Kesir. The field for miles was covered with the dead and dying. Those who were not dead were crying all night for help. We had fires built and caused a large number of them to return to them for it was a chilly night. The most of them were Irish and New York Zonaries. 1502 miles.

Saturday 28th

This morning presents a horrid sight, men mangled to pieces, dead and wounded are covering the ground for acres, baggage strewn in every direction, men calling for help, asking for water or desiring us to be the messenger to inform their parents and friends of their fate. It is a heart rendering scene. A true picture of war. Enough to make the bravest veteran weep. I hope I shall never again witness such a sight.

The wounded men would frequently ask what troops we were and when told that we were Jackson's men, the dying [Federal] soldiers would not believe us. They said that but a few evenings before it had been published to their troops that General Jackson had met with a signal defeat in the Valley and that he himself had been killed and his army routed, but when they fought us in the Valley, our guns were booming on their right wing.

Where our General would ride over the field, they would raise up as long as they could. I visited the hospitals and tried to relieve the suffering of the wounded as much as I could. The troops were busy buring the dead and removing the letters which were plentifully strewn over the field. Each was a Sanguine of success saying they would spend their Fourth of July in the Rebels capitol, but alas many of them lay lifeless upon the field or suffering from their wounds. They are to be petied.

Sunday 29th

Continued to bury the dead late in the evening we were ordered down to Woodbury bridge which crossed the Chickohomoney.[5] This had been destroyed by the enemy in his retreat. The engineers were then repairing it. General McGrider [sic, John B. Magruder] had engaged them at the York River Depot and completely routing them. Later we returned to camp. Miles 1505.

---

5 Chickohomoney. Refers to Chickahominy River.

Monday 30th

Crossed Chickohomoney and passed over the battle field of yesterday. The enemies dead were upon the ground by hundreds. They in their flight had abandoned their camp equipage, wagons and etc. Everything gave them the appearance of rout. Our march was through a continued encampment abandoned by the enemy. Longstreet had a warm engagement upon our right, completely routing them. We encamped at White Oak Swamp, a distance of 14 miles. 1519 miles.

May 1st - July, 1862

Crossed the White Oak Swamps passing through General Longstreet's battle field and soon came upon the enemy at Malvin Hill.[1] They drove our battery from their first position, they being entrenched upon a commanding hill. We were held in reserve until late in the evening and while in this position a stray shell mortally wounded Captain Fletcher and passing under my horse. Poor boy as noble and honest a fellow as ever sacrificed his life for liberty. He was carried to the rear and upon finding it mortal, I enformed him then of the nature of his wound. Said he: "I fear not to die. This is a good cause. Tell the boys to be in good spirits and do their duty." I then ask him if he had any message for his parents. "Tell them I fell honorable at my post in discharge of my duty."

It was too much for me. I bid him good bye and rejoined my regiment. This mishap seemed to fall like a pall upon the regiment. Never have I experienced anything in my life that had upon me the same effect.

It was dark when we were ordered into the field after crossing through swamps. We came into an open field. The enemy were shelling the field with all fury. The shells seemed to me to be bursting in the mens faces. We advanced upon them with no avail. We fell back to a church. The shelling continued for several hours after night. We returned and slept upon the field. 1529 miles.

---

1 Malvin Hill. Refers to Malvern Hill.

Wednesday 2nd

During the night the enemy retired leaving their dead and wounded. At one time last evening, I thought we were whipped, but fortune smiles upon us. Last night was one of the most horrifying things I ever witnessed. Everything seemed to be on fire. I had no idea of ever getting of the field again. When I went on, it seemed impossible for a person to escape, yet but a compartively [few] were killed. Our regiment lost 23 killed and wounded.

I visited the enemy's position this morning and I had no idea that they had half as strong a position as what they had. Three entrenchments encircled the hill comanding, in every respect, but our men did good execution. The countless dead spoke for them. We fell back some two miles and encamped. Rained all day. 1537 miles.

Thursday 3rd

Having finished burying the dead and collected a few of the stragglers, which I am sorry to say were many. I was determined to follow up the retreat. We took the Charles City Road to Linkey Bend where the enemy had retired under the protection of their gun boats.

Encamping along the river for some time, we were thrown in line of battle and remained in it until after dark and then falling back to a strip of woods in our rear. Captain Fletcher died today. He was as near to me as a brother. We had as officers associated together and he was so brave, young and hopeful. Poor boy! We will never forget him. 1551 miles.

Wednesday 4th

This is the day the enemy expected to be in Richmond. What a disappointment to them. Many of their letters to their friends told them, when they received this, the Rebel Capitol would be theirs. Many of them expected or rather feared we would evacuate it, and they now are thirty miles farther from it with a routed army and one third of it in our hands or dead upon the field. The rest hemmed in, in the space of a few miles along the James River. They celebrated the day with the booming of guns and music.

Friday 5th

Went on picket. The enemies pickets and ours became very sociable, exchanging papers and mingling with each other until the officers were compelled to stop it. Fell back to Turkey run, a distance of 4 miles. 1555 miles.

Saturday 7th

General Longstreet's division passed us today. The weather exceedingly hot.

Sunday 8th

Remained in camp until late in the afternoon and marched to White Oak Swamp. It being too hot for the men to march during the heat of the day. We did not reach camp until late at night. Very disagreeable marching. We passed several battle fields and the stench of dead men and horses was dreadful. Many of the men who had been buried were so that they could be seen. The dirt was removed. 1559 miles.

Monday 9th

Remained in camp until late in the evening and then marched near Mackameksville and encamped. I visited Richmond this evening late when we reached the city.

One could not imagine my feelings as I approached the city, having not been in a house since we left the valley and to see the ladies promenading, reminded me of better days. After riding around the best portion of the place, we went to Mr. Fisher's and remained the rest of the night. 1582 miles.

Tuesday 10th

Visited Hollywood[2] where the remains of President Monroe repose. Visited the war office and capital. The statues in the capital domain is grand. One will feel as if he were in the time of [Patrick] Henry when he looks upon the statue.

I could fancy that I heard his eloquence when he pronounced those patrick words in 76 -- "Liberty of Death."

In company with Dr. McGuire, I visited the theatre thinking it would carry my mind from the scenes of camp life.

Wednesday 11th

Returned to camp. Raining all day. Men are busy washing and cleaning up.

Friday 13th

Left camp this morning marching through Richmond after remaining here several hours. We took the train on the Central R.R.

You cannot imagine the feeling of the brigade when the once more were traveling in the direction of the Valley.

Our brigade in Richmond created considerable attraction. We encamped near the railroad bridge which crosses the Tapahanoe River, which had been destroyed by the enemy. Rained all night. Distance 8 miles. 1590 miles.

Saturday 16th

Took the train this morning and encamped near Louisa Court House. Only a portion of the brigade would be furnished transportation.

---

2 Hollywood. Hollywood Cemetery, where many famous statesmen and generals are buried.

Tuesday 18th

Took up the line of march for Gordonsville. Very disagreeable marching on account of the amount of rain which had fell in the past few days. We reached Gordonsville late in the evening passing through the town and encamping a mile beyond. 1606 miles.

Thursday 20th

Marched to camp McGruder, a very pleasant encampment. The enemy are some 14 miles from us at Culpepper Court House. General [John] Pope is in command and has issued his famous address to the army telling them that "He never saw the front of the foe." 1612 miles.

Friday 21st

Received orders to proceed at once to the Valley of Virginia to arrest deserters and conscripts. I did not fancy leaving the regiment at this time as the camp promised to be a very pleasant one. Commenced discharging men over 35 years of age.

Saturday 22nd

Started for the Valley arriving at the station at noon, reaching Staunton at sunset. I had put on my best and in camp I was considered the handsomest dressed gentleman among them, but on my arrival at Staunton, I was ashamed to make my appearance on the street. Weather very pleasant.

Monday 24th

As soon as my horse came up, I started for the valley. Visited Col. Harper and stayed the night at Mt. Crawford.

Tuesday 25th

After making all the necessary arrangements, I started for New Market.

Wednesday 26th

Reached Woodstock late in the evening. I was ordered back to Harrisonburg.

Friday 28th

Reached Harrisonburg and made preparation to open an office on the following day. I knew that my duty was an unpleasant one, yet I was determined to do my duty.

Saturday 29th

This morning I started the scouts throughout the country and arrested all the deserters in town and put them in the guard house ready to be sent off on the following morning.

August 1862

Wednesday 7th

Opened a conscript office in New Market. Left Lt. Williams in charge at Harrisonburg. We had sent off a large number of conscripts and deserters, creating a great excitement among the women by taking their husbands, fathers, brothers, sweethearts and sons. Without respect to persons and by my straightforward course, instead of making enemies, I made friends. Those who had relatives in the army thought it no more than just that the cowardly scoundrels who shirked the service ought to be compelled to defend his country.

About 10 o'clock P.M., it was reported that the enemy was advancing upon the place from Luray. When I awoke the town was in great confusion. Trunks were tumbling out of doors, buggies driving to and fro through the streets and the hasty riders could be heard galloping his horse, women and children were taking leave of their friends and relatives. Upon inquiry, I learned what was the occasion of this. My men had turned their horses loose in a field. I ordered them to gather their horses as soon as possible and started in direction of the mountain and found that the alarm was false. The enemy had crossed the river with a forrage train, but returning in the evening. I returned to the town and soon quieted the affairs of the excited populance, but by this time, some of them had almost reached Harrisonburg.

The turn pike was filled with wagons, cattle and horses. Farmers moving their stock off to prevent the Yankees from capturing it. For miles around the report had reached the ear of the farmers.

The excitement was even extended to Harrisonburg and many of the citizens made preparation to fly at the approach of the foe.

The excitement soon subsided and we were again wrapped in slumber. Our brigade which had been encamped at Liberty Mills when I left them, had fallen back to Green Springs. Pope[1] was understood advancing. General Jackson had received considerable reinforcements and surely they expected an engagement.

The enemy in Page Valley were committing serious depredations.

Thursday 8th

Established a provost guard and dispersed scouts throughout the neighborhood, which returned in the evening with a number of deserters and conscripts.

The delinquents are very much frightened and many have taken refuge in the mountains. We have scared a number to the army. The coaches to Staunton are crowded every evening.

The federals in Page are taking up a number of citizens in compliance with General Pope's order that where their men are fired upon, five loyal citizens are to be taken from the neighborhood as a hostage. In many instances, they have taken old men from a bed of illiness and dragged them to prison.

The mountains around Luray are filled with "Bush Whackers" who kill every federal they find. Quiet a number of them are daily killed by these men. They seem to have a great fear of them. I know one old gentleman who has killed 30 since they have been in the Valley. These "Bush Wackers" are of no small number. Many of these are a portion of the men that were cut of on our retreat up the Valley.

Nearly every state in the confederacy is represented in these hills and when they get a Yankee, they hoist the banner, "That dead men tell no tales."

---

1 Pope. Union General John Pope, b. 1822, grad. West Point, 1842. Achieved two brevets in Mexican War. In Virginia, lost the faith of his men and declared his headquarters was in the saddle. This prompted the quip that " he didn't know his headquarters from his hindquarters." His proposals on how to deal with the secessionists civilian population caused him to be the only Union commander to earn the personal animosity of General Robert E. Lee. Defeated at Cedar Mountain and Second Manassas.

These men are rendering good service and do more to cover the courage of the foe than a defeat in an open battle.

I wish every bush and rock in our Valley had a rifle behind it to pierce the heart of the ruthless foe. No news from the army today. We hourly expect the news of battle and fear not the issue.

Monday 11th

The news reached us today of the Cedar run fight, where the [Federal] army were shamefully routed with comparatively small loss upon our side. General Pope's proclamation to the army has not been correctly proven for instead of the foe running upon the sight of the union troops, they themselves fled.

This was a glorious fight upon the part of our brigade, reaping for herself fresh laurels, but while we boast of the brilliant achievement of our army, we must pause and drop a tear over our fallen braves. Among the number is General Chas. H. Winder, a brave and skillful officer. He was struck with a ball while placing his battery in position and soon expired. At first General Winder was not liked very much by the men of his command, but his coolness and judgement on battle field and his discipline in camp and upon the march, endeared him to every soldier. He was the most active man upon the field I ever knew.

Col. Fulkerson of the 2nd Brigade fell too, while leading his brigade in the charge. The enemies loss was heavy upon this occasion. The men who participated in it look upon it as fun for some of their regiments won without firing a single gun. We took a large number of prisoners. Our regiments loss was small. Col. Chas. Rowall of the 4th Va. Inf. commanded the brigade after the fall of Gen. Winder and with credit to himself. The brigade remained on the field during the night and next day fell back to their old encampment. The men were in high spirits over their success.

Tuesday 12th

The enemy evacuated Luray yesterday and marched to the assistance of Pope's Company with them a number of citizens as hostages. Several families informed me of the fact; and I immediately communicated the affair to General Jackson who had them released. They had almost demolished the town and the citizens felt as if they were not out of prison by having the restraints imposed upon them removed.

Deserters constantly arrived form the [Federal] army which I paroled and sent to Western Virginia. In many instances their pickets would desert their posts and come to our lines seeking paroles. These deserts were chiefly from the Western Virginia regiment furnished by Frank Pierpoint, the bogus Governor. There seems to be a great deal of demoralization in their army. General White still occupies Winchester and has fortified himself. A good amount of articles are constantly run through their lines.

We still find a number of conscripts and deserters in this portion of the Valley. Subsistance is scarce and high. It is with some difficulty that I can secure enough for my men. No news from our army except that reinforcements are constantly arriving. General [Braxton] Bragg is moving through Kentucky and everywhere meets with warm receptions by the citizens. We hope to hear of good work from the army of the Southwest.

Saturday 14th

I ordered Lt. Forestani to open an office in Woodstock, which he did. I then took charge of a scouting party and started for Newtown. The enemies line extended only to Kernstown, three miles this side of Winchester. When within a few miles of Newtown, I learned that they were in that place. In fact, we met a number of citizens leaving. We took such precautionary steps as we thought necessary and went into town, but no enemy was there. After placing out pickets, we had our horses fed and prepared for the night.

Sunday 15th

Yesterday morning Captain Rinker came to town with his company and we started on a scout to the Front Royal Road, where we understood that the enemy were passing with some trains. We visited the White Post, both up and down the road, but could discover nothing and late in the evening, we returned to town. It was reported to us that a band of thieves composed of Yankee soldiers and union men were committing some depredations upon the citizens near Taylors Furnace. After feeding the horses, we started cutting across the country to Neils Dam and took supper at Madison Campbells. Whereever we approached a house, the citizens would hide with fear thinking we were White's men who intended to impose the oath and when the mistake was discovered, they would extend to us all the hospitalities they could. We then took the mountain road and after traveling it for some time, we halted and encamped for the night.

Monday 16th

This morning we were informed by a "Bush Wacker" that some forty of a company to which they belonged had surrounded this party and intended to attack them at daylight and wishing our assistance.

After traveling some 8 miles, we reached the place at daylight. I desired to dismount and attack them, but Captain Rinker advocated a dash upon them mounted. After making every arrangement for the attack, we dashed upon them. They took shelter in three houses and concentrated their fire upon us, and nearly all of our men broke and ran, leaving the Captain, myself and five other men. We engaged the enemy until we had exhausted our ammunition. In the meantime, Captain Rinker was killed. Finding it impossible to withstand their fire and having no ammunition, we nearly surrendered. I ordered the men to draw their swords and we would cut through, but we met with no opposition. We arrived in Newtown late in the evening with the prisoners, losing 2 men and 3 horses, while the enemy had 3 killed and wounded. I started to return up the Valley. Reached Woodstock at midnight. The death of Captain Rinker a gallant officer, has been exceedingly sad news to the citizens of the Valley.

Tuesday 17th

My father and sister came through the lines. The enemy are ruling with an iron rule. Several cannon balls have been thrown into the town from their fortifications which have terrified the people.

I have had good success in the duties of collecting conscripts. We have sent off a large number. The army is still at Gordonsville.

A forward movement is daily expected. The enemy have evacuated Front Royal and it is generally believed that Pope is concentrating his forces at Culpepper.

Sunday 22nd

Opened an office in Luray. Lt. Jones in charge. The country is full of deserters. This is a pleasant town, noted for its wealth and beauty.

Monday 23rd

Returned to New Market ordered an office to be opened at Strausberg and Newtown.

Wednesday 25th

The enemy made a raid into Newtown and captured some of our scouts. It was owing to the neglect of the Lt. who failed to properly post sentinals. He was immediately placed in arrest and ordered to Woodstock.

Saturday 30th

Our army has advanced and every mail is looked for with great anxiety, expecting news of an engagement. We anticipated an early evacuation of Winchester.

September 1862

Tuesday 2nd

Reached Strausberg this evening and heard that we had whipped the enemy on the plains of Manassas. Several wounded men have arrived enforming us of the death of Col. Baylor, Neff and Barts. Sad news. The enemy advanced as far as Middletown, capturing several of my scouts. I think they intended to evacuate the place by making this demonstration, which is very probable if the news is true concerning the engagements at Manassas.

Having but few men and expecting the enemy to advance to cross the river and hold themselves in readiness to move at a moments notice. Wounded men continued to arrive, conferring report of a three day engagement at Manassas and that General Ewell lost his leg and General Jackson slightly wounded in the arm and that we had captured and destroyed a large amount of stores, totally destroying the train.

Wednesday 3rd

About 3 this morning a dreadful report was heard as if the noise of a distant gun, which was the explosion of the Magazine at Winchester. We knew they were evacuating the place and we were soon on the road to Winchester with all the force I could collect. Arrived there at 9 P.M. The streets were crowded with men, women, and children. They burnt a large portion of the town, leaving behind a large amount of commissary and quarter masters stores.

I sent a scouting party in to follow the enemy and commenced collecting the stores. Closed all the stores in town and placed guards all over the sutler stores, appointing Captain E. R. Smith Provost Marshall.

The town has been much abused by the enemy. One would scarcely know the place. The enemy had fallen back to Martinsburg 22 miles distant. I immediately communicated the facts to General Jackson by courier, he then being in Leesburg.

The citizens were very much rejoiced upon our return, but feared the enemy would return again. Picketed the town, but late in the evening, the enemy had advanced near the White House as of intending to return.

We had our horses saddeled all night, ready to move at any moment, while the best portion of the night was spent in examining letters which we had captured in the mail. We also captured several hundred prisoners which I immediately sent to Staunton.

Thursday 4th

Major Massie reported with his command from Woodstock. I set all the force I could command at work collecting all the stores and storing away. A large number of articles were carried off before I reached Winchester by the citizens. I searched many of the houses around town and found a number of valuable articles. We captured in all about 400 federal prisoners, the most of them I sent to Staunton. The rest were patroled and sent by New Creek. My duties were very laborious. Compelled to remain in my office the most of the night and hardly had time during the day to get my meals. The army had crossed into Maryland and were marching to Fredrick City. The enemy still occupying Harpers Ferry and Martinsburg.

Friday 5th

A number of stragglers came in from the army with some 200 sick. I established a hospital which was soon crowded. General Lee sent me the 17th Battalion of cavalry, which arrived today, commanded by Captain McDonal.

I received a dispatch this morning that the enemy were threatening the wagon train near Berryville. I dispatched the 17th Battallion to protect it and ordered the train to Millwood to await orders. Captain McDonal returned this evening reporting the enemy had fallen back to Charlestown.

We commenced reorganizing the stragglers which amounted to several hundred and put them into camp, distributing to them such arms as I could procure and placing them under the command of the best officers I could find.

Saturday 6th

Ordered Major Massie with three hundred cavalry to recover near Martinsburg. Nearly all of the wagon train from the army arrived here today with about 12,000 stragglers, making my position very responsible with but a few hundred cavalry to protect them with, but I now had organized nearly two regiments of stragglers and well armed and determined to protect the property as well as I could. Some 7 large hospitals were filled with sick, so I also established a field hospital, not having sufficient room in the town. Paroled some 50 yankees and sent them to Western Virginia.

General Lee has reached Fredrick City and his army is encamped around the city. The citizens have extended every kindness to our soldiers and welcomed them wherever they go. A number of them are destitute for clothing and now have a good opportunity of supplying themselves. Our money passed readily at par.

Sunday 7th

Major Massie advanced as far as Darksville, failing to take the necessary precaution and a few of the enemies cavalry dashed upon him, causing a rout, but few of the men were rallied to repulse them. Many of them, officers and men, came dashing into Winchester in full speed, some without hats or coats, reporting that the enemy had routed and killed nearly the whole of them.

I ordered the Provost Marshall to arrest them and he soon had the guard house filled with running cavalry. It created a dreadful panic among the citizens. They were running in every direction. All the wagons in town were soon galloping up the turnpike. I ordered all the force I could giving all the necessary directions to the Provost Marshall in case the enemy should continue to advance and we were unable to check them. All along the road was cowardly cavalry. I met Major Massie some 5 miles from town with but few men around him. The enemy had fallen back to Martinsburg. The enemies force only numbered 40. They captured 30 and killed one of our men. Such an act of cowardice I have not seen displayed in this war by either army.

I sent a scouting party under the command of Captain Griffin to follow the enemy and return to Winchester. By this time, the town had become quiet. Captain McDonal behaved himself with great credit and bravery but the men and a number of their officers have acted shamefully.

Stragglers still come in by the hundreds. We have some 4 regiments now organized. Many of them are without shoes and none can possibly be had. I have imprisoned all I can find which falls far short of meeting the demand. We have but few cooking untinsels. The men are compelled to bake their bread upon stones and cook their meat upon the end of their ramrods, yet with all these disadvantages, they are cheerful and willing to obey.

It is with great difficulty that we can get a sufficient supply of subsistance for the men we have here.

Monday 8th

The enemy threatened us today. I ordered the trains back on the roads thinking it safe, and also had all the combustable matter removed from the town. I sent a large number of prisoners off under a flag of truce to Romeny.

Tuesday 9th

Stragglers come in to the amount of 7,000 and committing various depredations among the citizens so much that I was compelled to establish guards at different parts of the county. Several men were shot while committing these depredations. The yankees had come into Aldie and the sick were brought to this place.

No preparations were made, not even cooking utinsels. After impressing the churches and every building in town, I had to have field hospitals.

Thursday 11th

Ex Governor [Enoch L.] Lowe of Maryland arrived this evening with an order from the President ordering me to send him to General Lee's headquarters immediately and without delay. General Lee was then in Fredrick City and I started him in a vehicle via Leesburg.

Friday 12th

Knowing that General Lee had entirely cut off the retreat of General White by Maryland and this being his only way to prevent being captured. I entertained some fears that this course would be the one he would adopt and not knowing the intention of the army, I at once thought I would take such steps as to secure the train and assist the army in Maryland, as it had been greatly reduced while I had some 12,000 of efficient men who would greatly aid them if pressed by McLelland who was moving out of Washington. I organized all the men into regiments and battallions and appointing officers to command. I found both officers and men willing to cooperate in the affair.

Arriving there the best I could, there were some two thousand barefooted men who I intended should remain there. From the broken down artillery, I selected 15 pieces, being well posted as to the number and strength of the enemy who had everything ready for the attack and upon the sight of us, he would run, but in case of a mistake, I got from the regiments already formed two battallions of volunteers, who I knew were trusty and if in case we would be pressed I would hold them in reserve and then hurl them upon the foe.

The artillery was under command of Major Richardson. I determined to attack the enemy in Martinsburg the following Sunday and immediately went to work to execute the plans. Issued ammunition, filled the kasons of the artillery, moving the regiments into camp, arranging them into three brigades and giving everything a general superintending myself.

The citizens had too become uneasy expecting us to fall into the hands of the foe and not depending upon the unorganized stragglers and in order to give them the confidence, I marched several of the best regiments through town, which had the desired effect. I then published to the troops an order telling them that the fate of us rested solely upon the strict adherence to their duty upon this occasion and that the country demanded it. The commissary department had supplies sufficient for expedition and several of the trains had been unloaded to give us transportation when night came.

I had seen everything arrainged with satisfaction and promised success.

Saturday 13th

This morning at 3, I received a dispatch from General Jackson ordering me to send him all the force of cavalry I could; that he had crossed the river and would attack them in the morning. This faded my hopes of success, bring the only time I ever had a chance of distinguishing myself.

White evacuated the place and retreated to Harpers Ferry. General ordered me to Martinsburg to take charge of the town. I started late in the evening arriving there at midnight. The enemy in his retreat had destroyed all the stores they could not carry away, which was enormous.

Sunday 15th

A portion of the army passed through here today on the way to the Ferry. After securing all the property I could, I left Lt. Jones in command and returned to Winchester. Canonading could be heard in direction of the Ferry and it was generally believed that the enemy would be compelled to surrender for capitulation was inevitable and every source of retreat was cut off.

Upon my return, I ordered the regiment of stragglers to be ready to move by daylight the next morning, having every arrangement made to supply them with provisions.

Stragglers are still coming in by the scores. Lt. Funk was ordered to report to me. Reached here today. The army is greatly reduced by stragglers. The country for a hundred miles is filled with them. While a large portion of the army is unfit for service for want of shoes, and yet with fatigue of the campaign, they are still cheerful and in good spirits. The armies of France never surpassed this.

One hundred yankee prisoners were paroled and sent off today to Romney.

Monday 16th

The enemy [at Harpers Ferry] surrendered yesterday to General Jackson with 11 thousand prisoners, 16,000 stand of small arms and 28 pieces of artillery with their trains and camp equipage. At first they attempted to retreat across the Shenendoah River, but General Hill had the Virginia Heights and General McLaw[1] had to Maryland Heights and wherever one of their batteries was placed into position, it was soon disabeled. As soon as our troops was placed in position, General Jackson ordered them to surrender. They refused, then the signal to attack was given and a concentrated fire was poured upon them soon causing them to hoist the white flag and our troops took position of the place.

---

1 General McLaw. Lafayette McLaws, b. Ga., 1821. grad. West Point, 1842. Served in Mexican War and expedition against the Mormons and fought the Navajos. Appointed Brig. Gen. 1861, Maj. Gen. 1862.

It was quiet a contrast between the two armies. The Federals were uniformed with much regularity and in the finest trim possible, while the Confederates marched triumphantly into the town, many barefooted and many even the officers were in such condition, none uniformed alike, this presenting a hideous appearance to their well dressed generals. Yet southern valor, even in rags, carried the day.

Colones Myles who was in command (though ranked by General White who declined) was mortally wounded and the unpleasant task of surrendering fell upon General White.

General Jackson with respect to rank and position declined receiving his sword. The officers were allowed to retain their side arms and private bagage and those who had horses were also permitted to keep them and each regiment a wagon to carry their baggage.

A large number of controbands were also captured and returned to their masters. A number of citizens from this portion of the Valley to their disgrace, were also captured. Their own feeling was enough to rebuke them without the contempt of the soldiers and the community.

General Hill was left in charge to send the baggage to the rear. General Jackson then crossed the river in direction of Boonsboro to assist General Longstreet, then pressed hard by McLelland. The engagement was seen from the Ferry, only a few miles distant.

September 16th, 1862

Ended the entries in this diary.

LATEST RELEASES & BEST SELLERS

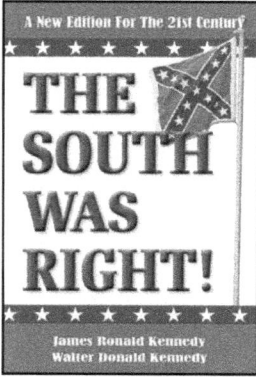

A New Edition For The 21st Century

THE SOUTH WAS RIGHT!

James Ronald Kennedy
Walter Donald Kennedy

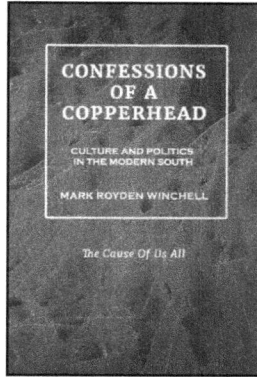

CONFESSIONS OF A COPPERHEAD

CULTURE AND POLITICS IN THE MODERN SOUTH

MARK ROYDEN WINCHELL

The Cause Of Us All

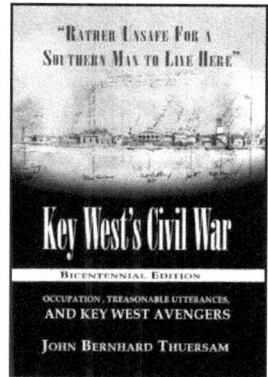

"RATHER UNSAFE FOR A SOUTHERN MAN TO LIVE HERE"

Key West's Civil War

BICENTENNIAL EDITION

OCCUPATION, TREASONABLE UTTERANCES, AND KEY WEST AVENGERS

JOHN BERNHARD THUERSAM

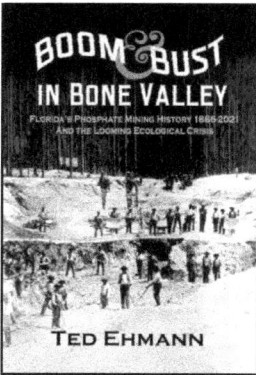

BOOM & BUST IN BONE VALLEY

FLORIDA'S PHOSPHATE MINING HISTORY 1886-2021 AND THE LOOMING ECOLOGICAL CRISIS

TED EHMANN

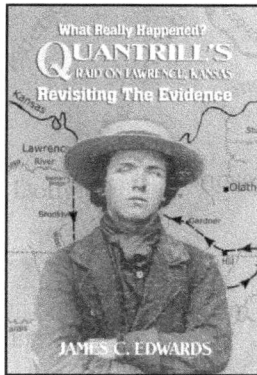

What Really Happened?
QUANTRILL'S RAID ON LAWRENCE, KANSAS
Revisiting The Evidence

JAMES C. EDWARDS

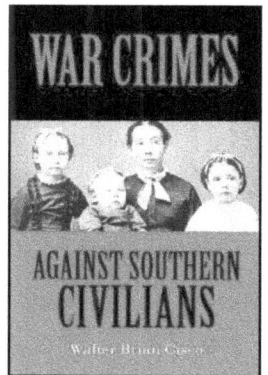

WAR CRIMES
AGAINST SOUTHERN CIVILIANS

Walter Brian Cisco

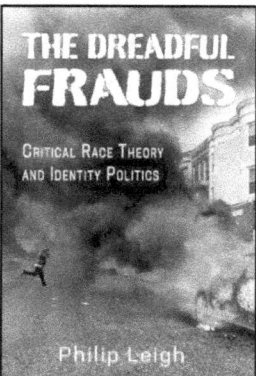

THE DREADFUL FRAUDS

CRITICAL RACE THEORY AND IDENTITY POLITICS

Philip Leigh

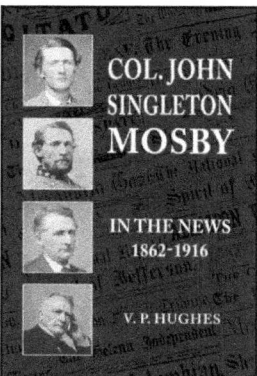

COL. JOHN SINGLETON MOSBY

IN THE NEWS 1862-1916

V. P. HUGHES

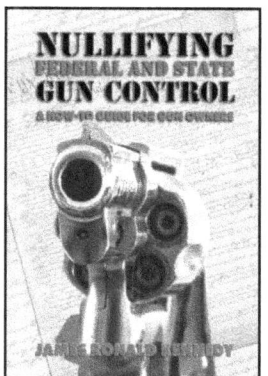

NULLIFYING FEDERAL AND STATE GUN CONTROL
A HOW-TO GUIDE FOR GUN OWNERS

JAMES RONALD KENNEDY

OVER 70 UNAPOLOGETIC UNRECONSTRUCTED TITLES FOR YOU TO ENJOY

SHOTWELLPUBLISHING.COM

Free Book Offer

Visit **FreeLiesBook.com**

Sign-up for new release notifications and receive a **FREE** downloadable edition of:

Lies My Teacher Told Me:
The True History of the War for
Southern Independence
by Dr. Clyde N. Wilson

and

Confederaphobia:
An American Epidemic
by Paul C. Graham

You can always unsubscribe and keep the book, so you've got nothing to lose!

www.ingramcontent.com/pod-product-compliance
Lightning Source LLC
Chambersburg PA
CBHW060052100426

42742CB00014B/2798